Differential Geometry and Group Theory

Alessio Zaccone

Differential Geometry and Group Theory

Springer

Alessio Zaccone
Dipartimento di Fisica
Università di Milano
Milano, Italy

ISBN 978-3-032-27517-2 ISBN 978-3-032-27518-9 (eBook)
https://doi.org/10.1007/978-3-032-27518-9

This Springer imprint is published by the registered company Springer Nature Switzerland AG
The registered company address is: Gewerbestrasse 11, 6330 Cham, Switzerland

In memoriam Kostya Trachenko

Foreword

Certain mathematical ideas appear so persistently throughout physics that, sooner or later, every physicist must come to terms with them. Differential geometry and Group theory undoubtedly belong to this category. Although these subjects originated within mathematics, they have become indispensable in the formulation of modern physics. From relativity to quantum theory, and from condensed matter to particle physics, the language of geometry and symmetry shapes the way we express and understand physical laws. Yet for all their importance, these subjects are not always presented to students in a manner that reflects their true role in physics. Too often they appear as formidable edifices of formalism, rather than as practical tools of physical reasoning.

My own appreciation of symmetry and geometry developed not through formal mathematical training but through physical problems. There was a period in my career when I became particularly interested in the role of symmetry in determining the physical properties of phases of matter. In those studies, I explored how the symmetry of an ordered phase dictates the form of its response functions. The natural language for such analysis is group theory and the decomposition of physical quantities into irreducible representations of the symmetry group. Once the symmetry is identified, a remarkable amount of physics follows almost automatically: the allowed tensor properties, the possible couplings between fields, and the structure of observable responses.

What struck me at the time was not merely the elegance of the mathematics, but the clarity it brought to physical reasoning. Symmetry arguments often reveal relationships that would otherwise remain buried under complicated calculations. They show that many physical properties are not accidental details of a particular model but necessary consequences of underlying invariances. In this sense, group theory and geometry are not simply mathematical tools; they are instruments of physical insight.

Yet there is a curious paradox. Although these ideas are central to modern physics, their presentation in many textbooks is often unnecessarily intimidating. Students encountering Differential geometry or Group theory for the first time frequently face a wall of formal definitions and abstract proofs, whose connection to physical intuition is not immediately apparent. The result is that many learners acquire a certain proficiency in symbolic manipulation but little sense of the geometric ideas that motivated the subject in the first place. Others abandon the subject altogether.

Physicists have long been aware of this difficulty. Progress in theoretical physics has almost always been guided by geometric intuition: the ability to visualize transformations, symmetries, and invariants. Einstein did not discover general relativity merely by manipulating tensor equations; he was guided by a deep geometric picture of spacetime. Likewise, in condensed matter physics, symmetry considerations routinely determine the physical properties of phases, defects, and collective modes. The mathematics involved can be sophisticated, but its real power lies in the insight it provides into the structure of physical phenomena.

The book by Alessio Zaccone addresses precisely this pedagogical challenge. It presents Differential geometry and Group theory with a clear emphasis on intuition and concrete calculation rather than excessive abstraction. The material is developed step by step, with intermediate stages of derivations carefully spelled out. Readers are not expected to accept results on faith; instead, they are guided through the reasoning that leads to them. This style of presentation is particularly valuable for students encountering these subjects for the first time, since it allows them to develop confidence in the mathematical tools while maintaining a clear connection to their physical meaning.

Equally important is the consistent effort to relate mathematical ideas to physical applications. Throughout the book, abstract concepts are illustrated by examples drawn from modern physics. Differential geometry naturally appears in the formulation of General Relativity and Field theory, while Group theory is connected to familiar topics such as rotations, angular momentum, crystallography, and the symmetry properties of solids. In this way, the reader is reminded that geometry and symmetry are not isolated mathematical curiosities but central elements of physical reasoning.

Another appealing aspect of the book is its effort to present differential geometry and group theory as parts of a unified conceptual framework. At first glance, these subjects may seem rather different: one deals with curved spaces and tensor calculus, the other with algebraic structures describing symmetry transformations. Yet both are fundamentally concerned with transformations and invariants. Geometry tells us how physical quantities behave when we change the coordinate frame used to describe them. Group theory classifies the transformations that leave certain physical properties unchanged. Seen from this perspective, the two subjects naturally belong together, and presenting them side by side highlights their deep conceptual connection.

The pedagogical philosophy underlying this book reflects an important principle: understanding in physics arises from the interplay between mathematical structure and physical intuition. Excessive formalism can obscure this relationship, while insufficient rigor can leave essential ideas vague. The challenge is to maintain a balance between clarity and precision. Zaccone's treatment manages to strike this balance by introducing the necessary mathematical tools while consistently emphasizing their physical meaning.

I am particularly pleased that Alessio Zaccone has undertaken the task of composing such a book. Alessio is well known for his work in theoretical condensed matter physics, particularly in mechanics of complex materials. In his research, he has repeatedly demonstrated a talent for combining rigorous theoretical analysis with clear physical intuition. It is therefore not surprising that the same qualities appear in this book. The exposition reflects the perspective of a practicing physicist who uses these mathematical methods as working tools rather than as abstract objects of study.

The book is aimed primarily at advanced undergraduate and early graduate students of physics, who often encounter these topics at the stage when the conceptual demands of modern theoretical physics begin to exceed the mathematical preparation offered in earlier courses. At the same time, it will be useful to many researchers who have acquired these methods in a somewhat fragmented way during their careers. Even experienced physicists occasionally benefit from revisiting familiar ideas in a clearer and more coherent exposition.

The importance of geometry and symmetry in physics can hardly be overstated. They provide a unifying language in which apparently unrelated phenomena can be understood within a common framework. From the curvature of spacetime to the symmetry classification of crystalline solids, from topological phases of matter to the representation theory underlying quantum mechanics, the same mathematical ideas reappear again and again. Books that illuminate these ideas while making them accessible to new generations of students perform an important service to the scientific community.

For these reasons, I am very pleased to recommend this book to its readers. It offers a thoughtful and carefully structured introduction to two subjects that lie at the heart of modern theoretical physics. Students approaching these topics for the first time will find here a guide that emphasizes understanding rather than intimidation, while more experienced readers may appreciate the clarity with which familiar concepts are revisited. In either case, the book provides a valuable pathway into the geometry and symmetry that underpin our understanding of the physical world.

Cavendish Laboratory
Cambridge, UK
2026

Eugene M. Terentjev

Preface

Differential geometry and group theory are the two pillars of geometry upon which the whole edifice of modern physics rests. Any serious theoretical physicist must be able to master these techniques, regardless of their field of specialization.

There are plenty of textbooks on mathematical methods for physics, and many textbooks on group theory and differential geometry. The vast majority of them are plagued by abstract formalism, which obscures the mental visualization of geometric concepts. In other words, most books available on this subject do not offer the kind of physical intuition without which the learner cannot achieve an in-depth, working understanding of tools that are essential for physical scientists at all levels. Historically, this state of affairs is largely a product of the supremacy of the abstract style in teaching mathematics introduced by the Bourbaki school around the middle of the last century. The contagion of the Bourbaki style has spread widely outside pure mathematics and has partly infected several schools of mathematical and theoretical physics.

This problem had already been stigmatized by Yakov B. Zeldovich, one of the greatest physicists in the second half of the twentieth century, who emphasized the damage that Bourbaki-style formalism can do to physics students. Captivated by the elegance of formalism, many of them are often unable to correctly visualize the underlying notions and objects, which are instead always connected, in some way, to our empirical representation of space. As emphasized by Zeldovich, for example in the prefaces to his own books of mathematical methods, *Higher Mathematics for Beginners* (with I.M. Yaglom) and *Elements of Applied Mathematics* (with A.D. Myskis), this tendency toward abstract formalism has catastrophic consequences for many (though of course not all) learners of theoretical physics and has produced legions of theorists with little physical intuition about the real world.

In the same spirit of Zeldovich, whose papers and books have greatly influenced my own work and approach to physics, the present textbook attempts to provide an anti-Bourbakist, accessible path for future scientists and engineers (including beginners) to penetrate the world of advanced geometry and its applications to physical problems.

Consistent with this program, the book provides a thorough, step-by-step introduction to tensor calculus, differential geometry (including its application to general relativity), the language of differential forms (with many examples from modern classical field theory), and group theory. This is done while systematically avoiding unnecessary abstract formalism and proofs, yet without sacrificing the necessary rigor.

What I hope will distinguish this book from many others on the market is that all intermediate steps in derivations and calculations are explicitly shown. The rich apparatus of worked-out calculation examples accompanying the presentation of concepts and derivations is an integral part of the learning process and is inseparable from the descriptive parts. Informal wording and explanations are deliberately favored over Bourbaki-style abstract language and formulae in order to guide the learner toward a clear visualization and understanding of the geometric objects and constructions. The attention to plain but precise language also helps connect the objects of geometry to our physical intuition and to our perception of the world.

I was much comforted when I read *The World as Will and Representation* by Schopenhauer and found, in the first book, the same views on geometry shared by Zeldovich and myself. As the German philosopher already pointed out two hundred years ago, the ultimate foundation of geometry is not logic but experience, that is, our empirically rooted intuition of space.

This book is based on a semester-long course that I taught for several years at the University of Milan. The course is addressed to final-year undergraduate students of physics and to first-year graduate students of physics, and it aims to help students bridge the gap between BSc-level courses and more specialized MSc-level courses of theoretical physics, nuclear physics, elementary particle physics (both theoretical and experimental), and condensed matter physics. The course was also often attended by mathematics students interested in a more intuitive treatment of these topics.

While primarily conceived for physics students and students of applied mathematics, this book may also help anyone, including the gifted amateur, who wishes to master the mathematics needed to understand the physics of the twentieth and twenty-first centuries. It may also be useful to engineering students, for example in aerospace engineering, structural engineering, and engineering physics. The systematic treatment of tensor calculus, diffeomorphisms, and calculus in curved spaces will also interest students of theoretical and computational solid mechanics, where symmetry concepts are now widely used, for example in metamaterials and tensegrity structures.

The book is suitable for a one-semester course in mathematical methods of physics at the final-year undergraduate or first-year graduate level. It can also serve as a resource for self-study and for researchers in the physical sciences who wish to fill gaps in their knowledge of modern physics, for which these tools are indispensable.

Students and researchers in solid-state and quantum physics will especially appreciate the connection with contemporary topological quantum physics, sometimes referred to as *quantum geometry*, including topics such as Berry curvature,

Chern numbers, the quantum Hall effect, and topological defects. Students and researchers in nuclear and particle physics will find the step-by-step treatment of the theory of angular momentum addition and Wigner matrices particularly useful.

Among the many application examples developed in the book, I should also mention general relativity, presented as a direct application of differential geometry, namely tensor calculus on curved manifolds. The corresponding chapter, together with the preceding ones that establish the necessary mathematical background, can also be used as a concise stand-alone introduction to general relativity, with all mathematical steps carefully explained and accompanied by physical intuition. Other applications discussed in the book include compatible and incompatible deformations in the theory of elasticity (relevant for solid mechanics), the covariant formulation of classical relativistic electrodynamics (also developed using differential forms), and the physical properties of crystals, including basic notions of crystallography and applications of group theory to infer ferroelectric and ferromagnetic properties.

One may wonder: why a book focused on two topics, differential geometry and group theory? The answer goes back to the programmatic declaration above. If the foundation of geometry lies in our empirical intuition of space, the most basic way we have to quantify it is by setting up a reference frame to measure distances between points and objects in space. In this neo-Schopenhauerian perspective, the reference frame—of which the Cartesian frame is the prototypical example—becomes the bridge between our intuition of space and geometry as a science.

From this point of view, differential geometry and group theory are both rooted in what might be called the art of transforming reference frames, and of moving objects from one frame to another. Tensor calculus and differential geometry provide tools that allow measurements to be represented in terms of functions of spatial points that are independent of the particular choice of reference frame. The connection with special relativity is evident and forms a leitmotif of the entire book.

If differential geometry is the art of representing measurements of physical quantities in spaces (manifolds) of arbitrary shape, group theory is the closely related art of representing transformations of objects in space that leave certain features unchanged. Here again the emphasis lies on transformations from one reference frame to another, as in the case of rotations.

Both differential geometry and group theory therefore place transformations of frames at the center of their reasoning, cultivating the ability to reason through transformations of perspective and to identify what is frame-independent. With these considerations in mind, it becomes natural to present differential geometry and group theory in a unified way that highlights the deep connections between these two branches of modern geometry.

Finally, this book is dedicated to the memory of my friend, collaborator, and fellow theoretical physicist Kostya Trachenko, Professor of physics at Queen Mary University of London, with whom I had the fortune to share several unforgettable intellectual adventures and a true friendship.

Milano, Italy
January 2026

Alessio Zaccone

Acknowledgments I would like to thank my colleagues in the Department of Physics of the University of Milan, in particular Prof. Stefano Forte and Prof. Alberto Pullia for offering me to teach a course in *Differential geometry and group theory* when I first joined the faculty in 2018. I am indebted to all the students who took this course at University of Milan, because, without their inquisitive attitude and clever questions, it would have been impossible to write this book. In particular, I would like to thank Lucrezia Bioni and Leonardo Cerasi for their help with the typesetting of my lectures. I am indebted to my academic mentors, in particular to Prof. Eugene Terentjev from whom I learned a lot in terms of tensor calculus, symmetry, and group theory.

Competing Interests The author has no competing interests to declare that are relevant to the content of this manuscript.

Contents

Chapter 1
Tensor Calculus

Abstract In this chapter we set the groundwork for the mathematical description of objects which can be classified based on their transformation laws under a change of coordinate system. We introduce the basic concepts of tensor calculus. We begin with coordinate transformations and the definition of tensors through their transformation laws. We then discuss the distinction between covariant and contravariant components, introduce the metric tensor, and study examples of coordinate systems frequently used in physics. These tools will form the mathematical foundation for the subsequent chapters. For more details and for additional examples the reader is referred to Cahill [1] and Riley et al. [2].

1.1 Coordinate Systems and Transformations Thereof

Physical laws can be formulated as equations involving quantities that remain invariant under changes of reference frame (RF). Such quantities are represented by tensors—mathematical objects that obey specific transformation rules when transitioning from one RF to another.

In general, tensors are spatial functions: they depend on ordered sets of spatial coordinates, which are experimentally measurable and therefore physical in nature, not merely mathematical abstractions. Spatial coordinates assign numerical labels to points of space (or spacetime). Different reference frames generally assign different coordinate values to the same physical point. Physical laws should therefore be formulated in a way that is independent of the particular coordinate system used to describe them. However, when described from different reference frames, that same point will be assigned different sets of coordinates, even though its existence is independent of any particular frame.

To make these ideas precise, we must specify what kinds of coordinate transformations are allowed. Only transformations that are smooth and invertible preserve the differentiable structure required to formulate physical laws.

A. Zaccone, *Differential Geometry and Group Theory*,
https://doi.org/10.1007/978-3-032-27518-9_1

Definition 1.1 Given a space, or a region thereof, consisting of points described by ordered n-tuples of continuous variables

$$\{x^k\}_{k=1,\dots,n} \subset \mathbb{R}, \quad n \in \mathbb{N},$$

an *admissible transformation* is defined as a set of functions

$$y^k = y^k(x^1, \dots, x^n), \quad k = 1, \dots, n,$$

satisfying the following conditions:

1. $y^k(x^1, \dots, x^n)$ is single-valued and continuously differentiable;
2. $\det\left[\frac{\partial y^k}{\partial x^j}\right] \neq 0$.

The condition on the Jacobian ensures that the transformation is locally invertible, allowing one to move consistently between coordinate systems.

1.2 Transformations of Tensors

We now introduce tensors as objects whose defining property is not their numerical values, but how those values change under admissible coordinate transformations.

Definition 1.2 A *tensor* of rank $r + s$ is a collection of quantities $T^{i_1 \dots i_r}_{j_1 \dots j_s}$ whose components depend on the coordinates and which transform under admissible coordinate transformations according to a specific multilinear rule (given below).

In order to express transformation laws compactly, we adopt the following index conventions:

1. **Einstein summation convention:** $A^k B_k \equiv \sum_{k=1}^{n} A^k B_k$;
2. **Range convention:** every free index in an equation independently ranges over $\{1, \dots, n\}$; thus, if an equation contains N free indices, it represents n^N distinct equations;
3. In $\frac{\partial y^k}{\partial x^j}$, the index j is considered a *subscript*.

Given a coordinate transformation, the components of a tensor in the new coordinate system are functions of the original components. This correspondence defines what is called an induced transformation.

Definition 1.3 Given an admissible coordinate transformation and a tensor, the *induced transformation* of the tensor's components is defined as

$$\tilde{T}^{i_1 \dots i_r}_{j_1 \dots j_s}(y^1, \dots, y^n) = \tilde{T}^{i_1 \dots i_r}_{j_1 \dots j_s}\left[T^{1 \dots 1}_{1 \dots 1}(x^1, \dots, x^n), \dots, T^{n \dots n}_{n \dots n}(x^1, \dots, x^n)\right].$$

Given a coordinate transformation $x \mapsto y$, the components of a tensor in the new coordinates are obtained from the old ones through a transformation law determined by the Jacobian of the coordinate transformation. This correspondence is called the *induced transformation*, which captures the idea that the same physical object is being described in different coordinate systems.

Theorem 1.1 *There exists an isomorphism between a coordinate transformation and its induced transformation.*

Proposition 1.1 *Every induced transformation is an isomorphism.*

These results formalize the intuitive expectation that changing coordinates does not alter the physical content of a tensor, but only its component representation.

In general, tensors are defined by their induced transformation laws, which may be linear or nonlinear (as in the case of curvilinear reference frames) but are always homogeneous. A prototypical example of a transformation law is the one between Cartesian and spherical coordinates in 3D,

$$\bar{x}^k = \bar{x}^k(x^1, x^2, x^3) \tag{1.1}$$

which we assume to be invertible ($\det J \neq 0$):

$$x^i = x^i(\bar{x}^1, \bar{x}^2, \bar{x}^3). \tag{1.2}$$

In this example we thus have:

$$\begin{aligned} x &= r \sin\theta \, \cos\varphi, \\ y &= r \sin\theta \, \sin\varphi, \\ z &= r \cos\theta. \end{aligned} \tag{1.3}$$

where (x, y, z) are the Cartesian coordinates and (r, θ, ϕ) are the spherical polar coordinates. Upon inversion, we obtain the formulae which express the curvilinear coordinates in terms of the Cartesian coordinates:

$$\begin{aligned} r &= \sqrt{x^2 + y^2 + z^2} \\ \theta &= \arccos \frac{z}{\sqrt{x^2 + y^2 + z^2}} \\ \varphi &= \operatorname{sgn}(y) \arccos \frac{x}{\sqrt{x^2 + y^2}}. \end{aligned} \tag{1.4}$$

This induced transformation is clearly a nonlinear one because trigonometric functions are involved.

In what follows, we will restrict our attention to tensors that are functions

$$\mathbb{R}^n \longrightarrow \mathbb{R}^{n^{r+s}}.$$

1.2.1 Absolute Tensors

We now classify tensors according to their transformation behavior, starting from the simplest cases.

Definition 1.4 A *scalar* is defined as a rank-0 tensor $\alpha(x^1, \ldots, x^n)$ with transformation law:

$$\tilde{\alpha}(y^1, \ldots, y^n) = \alpha(x^1, \ldots, x^n). \tag{1.5}$$

The value of a scalar field at a point in space is (rightly) independent of the reference frame (RF).

Definition 1.5 A *contravariant vector* is defined as a rank-1 tensor $a^k(x^1, \ldots, x^n)$ with transformation law:

$$\tilde{a}^k(y^1, \ldots, y^n) = \frac{\partial y^k}{\partial x^j} a^j(x^1, \ldots, x^n). \tag{1.6}$$

Definition 1.6 A *covariant vector* is defined as a rank-1 tensor $a_k(x^1, \ldots, x^n)$ with transformation law:

$$\tilde{a}_k(y^1, \ldots, y^n) = \frac{\partial x^j}{\partial y^k} a_j(x^1, \ldots, x^n). \tag{1.7}$$

To derive these transformation laws, one recalls the canonical isomorphism between a vector space V and its dual V^*: through this, a generic vector $\mathbf{v}$ can be expressed on the canonical basis $\{\mathbf{b}_i\}_{i=1,\ldots,n}$ of V as $\mathbf{v} = v^i \mathbf{b}_i$, or on the dual basis $\{\mathbf{b}^i\}_{i=1,\ldots,n}$ of V^* as $\mathbf{v} = v_i \mathbf{b}^i$.

In a Euclidean space with RF $(x^1, \ldots, x^n)$, the canonical basis at a point $\mathbf{r}$ is given by $\mathbf{e}_i = \frac{\partial \mathbf{r}}{\partial x^i}$. Given that $\mathbf{a} = a^j \mathbf{e}_j = \tilde{a}^k \tilde{\mathbf{e}}_k$, one finds:

$$\mathbf{e}_j = \frac{\partial \mathbf{r}}{\partial x^j} = \frac{\partial y^k}{\partial x^j} \frac{\partial \mathbf{r}}{\partial y^k} = \frac{\partial y^k}{\partial x^j} \tilde{\mathbf{e}}_k \quad \Longrightarrow \quad \tilde{a}^k = \frac{\partial y^k}{\partial x^j} a^j. \tag{1.8}$$

If instead $\mathbf{a}$ is expressed on the dual basis $\mathbf{e}^i = \nabla x^i$, then from $\mathbf{a} = a_j \mathbf{e}^j = \tilde{a}_k \tilde{\mathbf{e}}^k$ one obtains:

$$\mathbf{e}^j = \frac{\partial x^j}{\partial \mathbf{r}} = \frac{\partial x^j}{\partial y^k} \frac{\partial y^k}{\partial \mathbf{r}} = \frac{\partial x^j}{\partial y^k} \tilde{\mathbf{e}}^k \quad \Longrightarrow \quad \tilde{a}_k = \frac{\partial x^j}{\partial y^k} a_j. \tag{1.9}$$

The transformation law (1.7) coincides with that of the differentials dx^i, while (1.6) corresponds to the transformation law of the components of tangent vectors. These laws can be generalized to arbitrary tensors.

Definition 1.7 For a rank-$r+s$ tensor $T^{i_1\ldots i_r}_{j_1\ldots j_s}(x^1,\ldots,x^n)$, the transformation law is:

$$\tilde{T}^{i_1\ldots i_r}_{j_1\ldots j_s}(y^1,\ldots,y^n) = \frac{\partial y^{i_1}}{\partial x^{k_1}}\cdots\frac{\partial y^{i_r}}{\partial x^{k_r}}\frac{\partial x^{l_1}}{\partial y^{j_1}}\cdots\frac{\partial x^{l_s}}{\partial y^{j_s}}T^{k_1\ldots k_r}_{l_1\ldots l_s}(x^1,\ldots,x^n). \tag{1.10}$$

The Kronecker tensor, defined from the Kronecker delta, satisfies: $\tilde{\delta}^i_j = \frac{\partial y^i}{\partial x^k}\frac{\partial x^l}{\partial y^j}\delta^k_l = \frac{\partial y^i}{\partial y^j} = \delta^i_j$, and is thus isotropic (independent of RF).

The electric field, defined as the gradient of a scalar field (itself a rank-0 tensor), is $E_i = -\frac{\partial\phi}{\partial x^i}$.

A rank-2 tensor can be formed as the outer product of two rank-1 tensors, e.g. $\mathbf{T} = \mathbf{v}\otimes\mathbf{u}$, with contravariant, covariant, and mixed forms: $T^{jk} = v^j u^k, \quad T_{jk} = v_j u_k, \quad T^k_j = v_j u^k$.

The gradient of a vector is also a rank-2 tensor: $T^i_j = \frac{\partial v^i}{\partial x^j}$.

1.2.2 Pseudotensors

A particularly important class of coordinate transformations in $\mathbb{R}^3$ is given by rotations. These may preserve or reverse the orientation of space.

Definition 1.8 A *rotation* is an isometry of $\mathbb{R}^3$ represented by an orthogonal matrix

$$\mathbf{R} \in \mathrm{O}(3) \equiv \{\mathbf{R} \in \mathbb{R}^{3\times 3} : \mathbf{R}^\mathsf{T}\mathbf{R} = \mathbf{I}_3\}.$$

Definition 1.9 A *proper rotation* is a rotation represented by a matrix $\mathbf{R} \in \mathrm{SO}(3)$, i.e. $\det\mathbf{R} = +1$.

Definition 1.10 An *improper rotation* is a rotation represented by a matrix $\mathbf{R} \in \mathrm{O}(3)$ with $\det\mathbf{R} = -1$.

Improper rotations include reflections and therefore reverse the orientation of space.

Under general coordinate transformations, two distinct effects may occur: a change in orientation, associated with the sign of the determinant of the Jacobian, and a change in volume scale, associated with its absolute value. This distinction leads to different classes of tensorial objects.

In what follows, we focus on rotations in $\mathbb{R}^3$, where these effects can be cleanly separated.

Definition 1.11 A *pseudotensor* of rank $r+s$ is a tensorial object whose components transform, under a rotation, as

$$\tilde{T}^{i_1 \ldots i_r}_{j_1 \ldots j_s} = (\det \mathbf{R}) \frac{\partial y^{i_1}}{\partial x^{k_1}} \cdots \frac{\partial y^{i_r}}{\partial x^{k_r}} \frac{\partial x^{l_1}}{\partial y^{j_1}} \cdots \frac{\partial x^{l_s}}{\partial y^{j_s}} T^{k_1 \ldots k_r}_{l_1 \ldots l_s}. \tag{1.11}$$

A pseudotensor is therefore invariant under proper rotations but changes sign under improper rotations, reflecting its dependence on spatial orientation only.

The most important example of an orientation-sensitive object is the Levi-Civita symbol.

Definition 1.12 The *Levi-Civita symbol* in three dimensions is defined as

$$\epsilon_{ijk} = \begin{cases} +1 & (i, j, k) \text{ even permutation of } (1, 2, 3), \\ -1 & (i, j, k) \text{ odd permutation of } (1, 2, 3), \\ 0 & \text{otherwise.} \end{cases}$$

The following propositions can be proven.

Proposition 1.2 *Given* $\mathbf{A} \in \mathbb{R}^{3\times 3}$, *one has*

$$(\det \mathbf{A})\, \epsilon_{lmn} = A_{li} A_{mj} A_{nk}\, \epsilon^{ijk}.$$

Proposition 1.3

$$\epsilon_{ijk}\, \epsilon_{klm} = \delta_{il}\delta_{jm} - \delta_{im}\delta_{jl}.$$

From Proposition 1.2, the transformation law of the Levi-Civita symbol follows:

$$\tilde{\epsilon}_{ijk} = \frac{\partial x^l}{\partial y^i} \frac{\partial x^m}{\partial y^j} \frac{\partial x^n}{\partial y^k} \epsilon_{lmn} = (\det \mathbf{R})^{-1} \epsilon_{ijk} = (\det \mathbf{R})\, \epsilon_{ijk} \tag{1.12}$$

Thus, under improper rotations, the Levi-Civita symbol changes sign. However, under a general coordinate transformation, the Levi-Civita symbol acquires the full determinant factor of the Jacobian, not merely its sign. It therefore transforms as a *tensor density* of weight $w = -1$, rather than as a pseudotensor (which corresponds to $w = 0$). More precisely, the Levi-Civita symbol is a pseudotensor density: it changes sign under orientation reversal and also carries density weight -1. This general form unifies absolute tensors, pseudotensors, and tensor densities through the weight parameter w.

Pseudotensors acquire a sign change under orientation-reversing transformations, whereas tensor densities acquire a power of the Jacobian determinant and therefore also encode changes of volume scale. The Levi-Civita symbol is thus best understood as a tensor density: its sign change under improper rotations reflects orientation dependence, while the determinant factor encodes volume scaling.

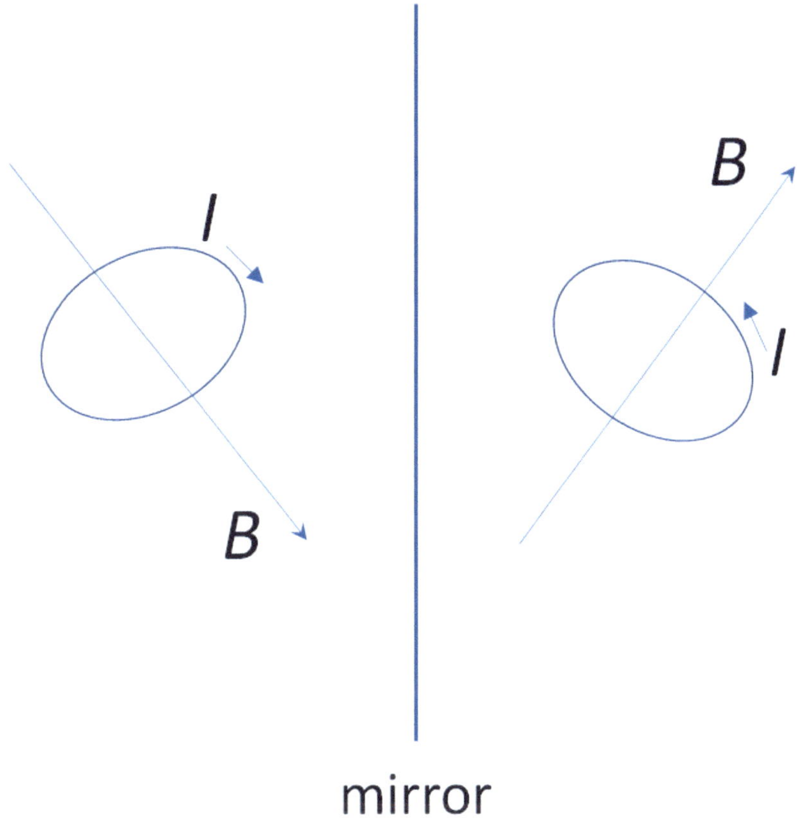

Fig. 1.1 The magnetic field **B** is a pseudovector because of its definition in terms of the Levi-Civita symbol. Under an improper rotation, such as a mirror reflection, not only it is reflected through the mirror but it also changes sign. In simple words, the vector, in addition to being reflected by the mirror, is also flipped

The magnetic field is defined as $B^i = (\nabla \times \mathbf{A})^i = \epsilon^{ijk} \frac{\partial A_k}{\partial x^j}$. We see that, due to the presence of the Levi-Civita symbol in its definition, under improper rotations such as reflections, the magnetic field changes sign. The situation is depicted in Fig. 1.1.

1.3 Metric Tensor

While in the Euclidean case the canonical basis $\mathbf{e}_j = \frac{\partial \mathbf{r}}{\partial x^j}$ ($\mathbf{e}_j$ tangent to the j-th coordinate line) coincides with the dual basis $\mathbf{e}^j = \nabla x^j$ ($\mathbf{e}^j$ orthogonal to the j-th coordinate line), this is not generally true.

In general Riemannian spaces, to go from a basis to its dual it is necessary to introduce a tensorial object called the metric tensor.

To measure lengths and angles in a coordinate system one must introduce an additional geometric structure: the metric tensor.

Definition 1.13 Given a basis $\{\mathbf{e}_j\}_{j=1,\dots,n}$, the *metric tensor* in that basis is defined as $g_{ij} \equiv \mathbf{e}_i \cdot \mathbf{e}_j \ \forall i, j = 1, \dots, n$.

Obviously, one can also specify the contravariant components g^{ij} instead of the covariant ones.

If the basis vectors are mutually orthogonal, the metric tensor will be represented by a diagonal matrix: just as in vector spaces one can always obtain an orthonormal basis from a non-orthonormal one (Gram-Schmidt algorithm), the metric tensor can also be diagonalized.

Proposition 1.4 $ds^2 = g_{ij} dx^i dx^j$.

Proof $ds^2 \equiv d\mathbf{r} \cdot d\mathbf{r} = dx^i \mathbf{e}_i \cdot dx^j \mathbf{e}_j = g_{ij} dx^i dx^j$. □

Proposition 1.5 *In the three-dimensional case,* $dV = \sqrt{g}\, dx^1 dx^2 dx^3$, *with* $g \equiv \det g_{ij}$.

Proof Considering an already orthogonalized basis such that $g_{ij} = h_i^2 \delta_{ij}$, where h_i is called the scale factor, we have $ds^2 = h_i^2 (dx^i)^2$ (consistent with $d\mathbf{r} = \frac{\partial \mathbf{r}}{\partial x^j} dx^j = h_j dx^j \hat{\mathbf{e}}_j$ by definition of $\hat{\mathbf{e}}_j$); by definition $dV = |dx^1 \mathbf{e}_1 \cdot (dx^2 \mathbf{e}_2 \times dx^3 \mathbf{e}_3)| = h_1 h_2 h_3 dx^1 dx^2 dx^3 \hat{\mathbf{e}}_1 \cdot (\hat{\mathbf{e}}_2 \times \hat{\mathbf{e}}_3)$, but this basis is orthonormal and for the metric tensor $\det \mathbf{g} = h_1^2 h_2^2 h_3^2$, thus $dV = \sqrt{g}\, dx^1 dx^2 dx^3$. □

This proposition can be generalized to the n-dimensional non-orthogonal case (for the definition of the wedge product, see Sect. 3.1).

Proposition 1.6 *In an n-dimensional manifold with metric tensor* g_{ij}:

$$dV = \sqrt{g} \bigwedge_{k=1}^{n} dx^k \tag{1.13}$$

The metric tensor is thus fundamental for the calculation of distances, areas, volumes, etc., but it also has another important role: converting between covariant and contravariant components.

The scalar product can be expressed in various ways: $\mathbf{a} \cdot \mathbf{b} = a_i b^i = a^i b_i = g_{ij} a^i b^j = g^{ij} a_i b_j$, from which two important identities follow:

$$\begin{aligned} g_{ij} a^i &= a_j \\ g^{ij} a_i &= a^j \end{aligned} \tag{1.14}$$

It is thus possible to go from a basis to its dual via contraction with the metric tensor.

Proposition 1.7 $\left[g^{ij}\right] = \left[g_{ij}\right]^{-1}$.

Proof $a^i = g^{ij} a_j = g^{ij} g_{jk} a^k$, hence $g^{ij} g_{jk} = \delta^i_k$, i.e., $[g^{ij}][g_{ij}] = \mathbf{I}_n$. □

From this proof we also note that $g^i_j = \delta^i_j$.

Given the relation between covariant and contravariant components, one can state in general the transformation law of a tensor.

Definition 1.14 Given a tensor $T^{i_1 \ldots i_r}_{j_1 \ldots j_s}(x^1, \ldots, x^n)$ of rank $r+s$, its transformation law is:

$$\tilde{T}^{i_1 \ldots i_r}_{j_1 \ldots j_s}(y^1, \ldots, y^n) = \det \left[\frac{\partial x}{\partial y} \right]^w \frac{\partial y^{i_1}}{\partial x^{k_1}} \cdots \frac{\partial y^{i_r}}{\partial x^{k_r}} \frac{\partial x^{l_1}}{\partial y^{j_1}} \cdots \frac{\partial x^{l_s}}{\partial y^{j_s}} T^{k_1 \ldots k_r}_{l_1 \ldots l_s}(x^1, \ldots, x^n) \tag{1.15}$$

where depending on the value of w one speaks of:

- $w = 0$: ordinary (absolute) tensor;
- $w \neq 0$: tensor density of weight w.

Pseudotensors are not characterized by a density weight. Instead, they acquire an additional factor sgn(det J) under orientation-reversing transformations. Tensor densities and pseudotensors are distinct concepts, although an object may possess both properties.

Note that $\det \left[\frac{\partial x}{\partial y} \right] = \frac{1}{\det \mathbf{J}}$, where $\mathbf{J}$ is the Jacobian of the transformation.

Theorem 1.2 (Quotient Theorem) *Given two tensors* **A** *and* **C**, *if for an object* **B** *the tensorial equation* **AB** = **C** *holds, then* **B** *is a tensor.*

1.4 Moving Reference Frames

It is possible to consider reference frames in which the basis vectors depend on the point where they are applied: given a point $\mathbf{r}(\mathbf{x})$, the infinitesimal displacement from it is written as $d\mathbf{r}(\mathbf{x}) = \mathbf{e}_j(\mathbf{x}) dx^j$, and the basis vectors are $\mathbf{e}_j(\mathbf{x}) = \frac{\partial \mathbf{r}}{\partial x^j}(\mathbf{x})$.

When moving from one reference frame to another, they transform as covariant vectors:

$$\tilde{\mathbf{e}}_j(\mathbf{y}) = \frac{\partial \mathbf{r}}{\partial y^j} = \frac{\partial x^k}{\partial y^j} \frac{\partial \mathbf{r}}{\partial x^k} = \frac{\partial x^k}{\partial y^j} \mathbf{e}_k(\mathbf{x}). \tag{1.16}$$

These are also vectors in the embedding manifold (a higher-dimensional manifold, usually Euclidean, that "covers" the considered manifold; for example, $\mathbb{S}^2$ is embedded in $\mathbb{R}^3$), which has a metric tensor η_{ab}.

These vectors can be both non-normalized and non-symmetric; moreover, they define a metric tensor on the embedding manifold of dimension n:

$$\mathbf{e}_i(\mathbf{x}) \cdot \mathbf{e}_j(\mathbf{x}) = \sum_{a,b=0}^{n} \mathbf{e}_i{}^a(\mathbf{x})\, \eta_{ab}\, \mathbf{e}_j{}^b(\mathbf{x}), \qquad \mathbf{e}_i{}^a(\mathbf{x}) \equiv \frac{\partial X^a(\mathbf{x})}{\partial x^i}. \tag{1.17}$$

The map $X^a(\mathbf{x})$ is the **embedding**. The dot denotes the ambient inner product with metric η_{ab}. Consequently, $g_{ij}(\mathbf{x}) = \eta_{ab}\, \partial_i X^a\, \partial_j X^b$ is the **induced metric** on the sub-manifold.

In general, the intrinsic coordinates x and ambient coordinates X are different objects related by the embedding $X^a(x)$.

For real manifolds, the metric tensor is always real and symmetric.

Under a transformation from one moving frame to another, the metric tensor transforms as a doubly covariant tensor:

$$\tilde{g}_{ij}(\mathbf{y}) = \tilde{\mathbf{e}}_i(\mathbf{y}) \cdot \tilde{\mathbf{e}}_j(\mathbf{y}) = \frac{\partial x^k}{\partial y^i}\mathbf{e}_k(\mathbf{x}) \cdot \frac{\partial x^l}{\partial y^j}\mathbf{e}_l(\mathbf{x}) = \frac{\partial x^k}{\partial y^i}\frac{\partial x^l}{\partial y^j} g_{kl}(\mathbf{x}). \tag{1.18}$$

In the case of orthogonal systems, one can compute the scale factors $h_i \equiv \|\mathbf{e}_i(\mathbf{x})\|$, in order to orthonormalize the system: $\hat{\mathbf{e}}_i = \frac{\mathbf{e}_i}{h_i}$. In this case, $g_{ij} = h_i^2 \delta_{ij}$.

Proposition 1.8 $\hat{\mathbf{e}}_i = h_i \mathbf{e}^i$.

Proof Note that no sum over i is implied; in this case:

$$\hat{\mathbf{e}}_i = \frac{1}{h_i}\mathbf{e}_i = \frac{1}{h_i} g_{ij}\mathbf{e}^j = \frac{1}{h_i} h_i^2 \delta_{ij}\mathbf{e}^j = h_i \mathbf{e}^i \tag{1.19}$$

□

1.4.1 Example: The Sphere S^2

Let the point $\mathbf{P}$ be a 3-dimensional Euclidean vector that represents a point on the spherical surface of radius r (a 2-dimensional manifold). The spherical coordinates (θ, φ) are defined so that $\mathbf{P} = \mathbf{P}(\theta, \varphi)$.

The basis vectors are

$$\mathbf{e}_\theta = \frac{\partial \mathbf{P}}{\partial \theta}, \qquad |\mathbf{e}_\theta| = r, \tag{1.20}$$

$$\mathbf{e}_\varphi = \frac{\partial \mathbf{P}}{\partial \varphi}, \qquad |\mathbf{e}_\varphi| = r \sin\theta. \tag{1.21}$$

Hence, the scale factors are:

$$h_\theta = |\mathbf{e}_\theta| = r, \tag{1.22}$$

$$h_\phi = |\mathbf{e}_\varphi| = r \sin\theta. \tag{1.23}$$

The scalar products give the components

$$g_{ij}(\mathbf{x}) = \mathbf{e}_i(\mathbf{x}) \cdot \mathbf{e}_j(\mathbf{x}). \tag{1.24}$$

Thus, the metric tensor of the sphere is given by the matrix

$$(g_{ij}) = \begin{pmatrix} g_{\theta\theta} & g_{\theta\varphi} \\ g_{\varphi\theta} & g_{\varphi\varphi} \end{pmatrix} = \begin{pmatrix} \mathbf{e}_\theta \cdot \mathbf{e}_\theta & \mathbf{e}_\theta \cdot \mathbf{e}_\varphi \\ \mathbf{e}_\varphi \cdot \mathbf{e}_\theta & \mathbf{e}_\varphi \cdot \mathbf{e}_\varphi \end{pmatrix} \tag{1.25}$$

$$= \begin{pmatrix} r^2 & 0 \\ 0 & r^2 \sin^2\theta \end{pmatrix}. \tag{1.26}$$

The determinant of the metric is

$$\mathrm{g} \equiv \det(g_{ij}) = r^4 \sin^2\theta. \tag{1.27}$$

1.4.2 Remarks

The points are physical quantities independent of the chosen coordinates. The distance $|\mathbf{P} - \mathbf{Q}|$ between two points $\mathbf{P}$, $\mathbf{Q}$ is an invariant quantity. If $\mathbf{Q} = \mathbf{P} + d\mathbf{P}$, then the two points lie on a manifold (space-time) and are infinitesimally close. The displacement vector is written as

$$d\mathbf{P} = \mathbf{e}_i dx^i. \tag{1.28}$$

1.4.2.1 Quadratic Form of the Displacement

The infinitesimal displacement $d\mathbf{P}$ is a sum of basis vectors multiplied by coordinate differentials dx^i. Its scalar product with itself gives the quadratic form of the displacement, which is invariant under a change of coordinates.

$$d\mathbf{P}^2 = d\mathbf{P} \cdot d\mathbf{P} = (\mathbf{e}_i dx^i) \cdot (\mathbf{e}_j dx^j) \tag{1.29}$$

$$= g_{ij} dx^i dx^j. \tag{1.30}$$

Thus, in another coordinate system, the same expression holds:

$$d\mathbf{P}^2 = (\mathbf{e}'_i dx'^i) \cdot (\mathbf{e}'_j dx'^j) \tag{1.31}$$

$$= g'_{ij} dx'^i dx'^j. \tag{1.32}$$

This invariance is consistent with the rule of Gaussian coordinates: the quadratic form $d\mathbf{P}^2$ is a second-order invariant tangent to the manifold.

1.4.2.2 Orthonormal Basis Vectors

The orthonormal basis vectors $\hat{\mathbf{e}}_i$ are obtained from the coordinate basis vectors $\mathbf{e}_i$ by normalizing them, i.e. dividing by the scale factors h_i that characterize the orthogonal coordinate system:

$$\hat{\mathbf{e}}_i = \frac{\mathbf{e}_i}{h_i} = h_i\, \mathbf{e}^i, \qquad \hat{\mathbf{e}}_i \cdot \hat{\mathbf{e}}_j = \delta_{ij}. \tag{1.33}$$

Here $\mathbf{e}^i$ is the dual basis vector, reciprocal to $\mathbf{e}_i$, and h_i is the multiplicative scale factor.

A physical vector $\mathbf{V}$ can be written as:

$$\mathbf{V} = \mathbf{e}_i V^i = h_i \hat{\mathbf{e}}_i V^i = \hat{\mathbf{e}}_i\, (h_i V^i) \tag{1.34}$$

$$= \hat{\mathbf{e}}_i V_i, \tag{1.35}$$

where $V_i = h_i V^i$ are the physical components of $\mathbf{V}$ with respect to the orthonormal basis.

1.4.2.3 Scalar Product and Polar Coordinates

The components of a vector in the orthonormal basis are

$$V_i = h_i V^i, \tag{1.36}$$

where there is no implicit summation over i.

The scalar product between two invariant vectors is then

$$\mathbf{V} \cdot \mathbf{U} = g_{ij} V^i U^j = V_i U^i. \tag{1.37}$$

1.5 Example: Polar Coordinates in 2D

Consider the plane (x, y) in polar coordinates (r, θ). The displacement $d\mathbf{p}$ corresponding to a variation of coordinates is

$$d\mathbf{p} = \hat{\mathbf{e}}_r\, dr + \hat{\mathbf{e}}_\theta\, r\, d\theta. \tag{1.38}$$

Here the basis vectors are:

$$\mathbf{e}_r = \hat{\mathbf{e}}_r, \tag{1.39}$$

$$\mathbf{e}_\theta = r\, \hat{\mathbf{e}}_\theta. \tag{1.40}$$

The metric tensor in polar coordinates is:

$$(g_{ij}) = \begin{pmatrix} 1 & 0 \\ 0 & r^2 \end{pmatrix}. \tag{1.41}$$

1.5.1 Contravariant Basis Vectors

The contravariant basis vectors are:

$$\mathbf{e}^r = \hat{\mathbf{e}}_r, \tag{1.42}$$

$$\mathbf{e}^\theta = \frac{1}{r}\hat{\mathbf{e}}_\theta. \tag{1.43}$$

with the respective scale factors given by:

$$h_r = 1, \tag{1.44}$$

$$h_\theta = r. \tag{1.45}$$

1.5.2 Physical Vector in Polar Coordinates in 2D

Thus a physical vector $\mathbf{V}$ is written as:

$$\mathbf{V} = V^i \mathbf{e}_i = V_r \hat{\mathbf{e}}_r + V_\theta \hat{\mathbf{e}}_\theta, \tag{1.46}$$

where V_r and V_θ are the components with respect to the orthonormal basis.

1.6 Example: Cylindrical Coordinates in 3D

In cylindrical coordinates in three-dimensional space, the infinitesimal displacement $d\mathbf{p}$ at a point P is given by the variation of its coordinates:

$$d\mathbf{p} = \frac{\partial \mathbf{p}}{\partial \rho} d\rho + \frac{\partial \mathbf{p}}{\partial \varphi} d\varphi + \frac{\partial \mathbf{p}}{\partial z} dz \tag{1.47}$$

$$= \mathbf{e}_\rho \, d\rho + \mathbf{e}_\varphi \, d\varphi + \mathbf{e}_z \, dz. \tag{1.48}$$

The basis vectors are:

$$\mathbf{e}_\rho = \hat{\mathbf{e}}_\rho, \qquad \mathbf{e}_\varphi = \rho \, \hat{\mathbf{e}}_\varphi, \qquad \mathbf{e}_z = \hat{\mathbf{e}}_z, \tag{1.49}$$

where $\hat{\mathbf{e}}_\rho, \hat{\mathbf{e}}_\varphi, \hat{\mathbf{e}}_z$ are unit vectors. Thus, the scale factors are:

$$h_\rho = 1, \qquad h_\varphi = \rho, \qquad h_z = 1.$$

The metric tensor in cylindrical coordinates is:

$$(g_{ij}) = (\mathbf{e}_i \cdot \mathbf{e}_j) = \begin{pmatrix} 1 & 0 & 0 \\ 0 & \rho^2 & 0 \\ 0 & 0 & 1 \end{pmatrix}. \tag{1.50}$$

Its determinant is

$$g = \det(g_{ij}) = \rho^2. \tag{1.51}$$

The invariant volume element is given by:

$$dV = \sqrt{g}\, dx^1 dx^2 dx^3 \tag{1.52}$$

$$= \rho\, d\rho\, d\varphi\, dz. \tag{1.53}$$

1.6.1 Contravariant Basis Vectors

The contravariant basis vectors are:

$$\mathbf{e}^\rho = \hat{\mathbf{e}}_\rho, \qquad \mathbf{e}^\varphi = \frac{1}{\rho}\hat{\mathbf{e}}_\varphi, \qquad \mathbf{e}^z = \hat{\mathbf{e}}_z, \tag{1.54}$$

with the respective scale factors:

$$h_\rho = 1, \qquad h_\varphi = \rho, \qquad h_z = 1. \tag{1.55}$$

Note that these scale factors refer to the coordinate (covariant) basis vectors, even though they appear in the subsection about contravariant basis vectors. They are reused there because the contravariant basis is constructed from their inverses.

1.6.2 Vector in Cylindrical Coordinates

A physical vector $\mathbf{V}$ is written as:

$$\mathbf{V} = V^i \mathbf{e}_i = V_\rho\, \hat{\mathbf{e}}_\rho + V_\varphi\, \hat{\mathbf{e}}_\varphi + V_z\, \hat{\mathbf{e}}_z. \tag{1.56}$$

The unit vectors in cylindrical coordinates can be expressed in terms of Cartesian unit vectors:

$$\hat{\mathbf{e}}_\rho = \cos\varphi\,\hat{\mathbf{e}}_x + \sin\varphi\,\hat{\mathbf{e}}_y, \tag{1.57}$$

$$\hat{\mathbf{e}}_\varphi = -\sin\varphi\,\hat{\mathbf{e}}_x + \cos\varphi\,\hat{\mathbf{e}}_y, \tag{1.58}$$

$$\hat{\mathbf{e}}_z = \hat{\mathbf{e}}_z. \tag{1.59}$$

In this way, the position vector is

$$\mathbf{r} = (\rho\cos\varphi,\ \rho\sin\varphi,\ z). \tag{1.60}$$

1.7 Example: Spherical Coordinates in 3D

In spherical coordinates (r, θ, φ), the infinitesimal displacement is:

$$d\mathbf{p} = \hat{\mathbf{e}}_r\,dr + \hat{\mathbf{e}}_\theta\,r\,d\theta + \hat{\mathbf{e}}_\varphi\,r\sin\theta\,d\varphi. \tag{1.61}$$

Thus basis vectors are:

$$\mathbf{e}_r = \hat{\mathbf{e}}_r, \tag{1.62}$$

$$\mathbf{e}_\theta = r\,\hat{\mathbf{e}}_\theta, \tag{1.63}$$

$$\mathbf{e}_\varphi = r\sin\theta\,\hat{\mathbf{e}}_\varphi. \tag{1.64}$$

The metric tensor is:

$$(g_{ij}) = \begin{pmatrix} 1 & 0 & 0 \\ 0 & r^2 & 0 \\ 0 & 0 & r^2\sin^2\theta \end{pmatrix}. \tag{1.65}$$

The determinant is

$$g = \det(g_{ij}) = r^4\sin^2\theta. \tag{1.66}$$

The invariant volume element is then:

$$dV = \sqrt{g}\,dx^1dx^2dx^3 \tag{1.67}$$

$$= r^2\sin\theta\,dr\,d\theta\,d\varphi. \tag{1.68}$$

1.7.1 Orthonormal and Contravariant Bases

The orthonormal basis vectors in spherical coordinates are:

$$\hat{\mathbf{e}}_r = \hat{\mathbf{r}}, \qquad \hat{\mathbf{e}}_\theta = \hat{\boldsymbol{\theta}}, \qquad \hat{\mathbf{e}}_\varphi = \hat{\boldsymbol{\varphi}}. \tag{1.69}$$

The contravariant basis vectors are:

$$\mathbf{e}^r = \hat{\mathbf{e}}_r, \qquad \mathbf{e}^\theta = \frac{1}{r}\hat{\mathbf{e}}_\theta, \qquad \mathbf{e}^\varphi = \frac{1}{r\sin\theta}\hat{\mathbf{e}}_\varphi. \tag{1.70}$$

A physical vector $\mathbf{V}$ is written as:

$$\mathbf{V} = V^i\mathbf{e}_i = V_r\hat{\mathbf{e}}_r + V_\theta\hat{\mathbf{e}}_\theta + V_\varphi\hat{\mathbf{e}}_\varphi. \tag{1.71}$$

1.8 Example: Gradient of a Scalar Field in Curvilinear Coordinates

Let $f(x)$ be a scalar field. The differential of f is given by:

$$df(x) = f(x+dx) - f(x) = \frac{\partial f}{\partial x^i}dx^i. \tag{1.72}$$

Since

$$d\mathbf{p} = \mathbf{e}_i dx^i, \tag{1.73}$$

we can write in invariant form:

$$df = \nabla f \cdot d\mathbf{p}. \tag{1.74}$$

The gradient is defined as:

$$\nabla f = \hat{\mathbf{e}}_i \frac{1}{h_i}\frac{\partial f}{\partial x^i}, \tag{1.75}$$

where h_i are the scale factors.

Thus,

$$\nabla f \cdot d\mathbf{p} = \frac{\partial f}{\partial x^i}dx^i = df. \tag{1.76}$$

1.8.1 Gradient in 2D Polar Coordinates

In polar coordinates, we have:

$$\nabla f = \hat{\mathbf{e}}_r \frac{\partial f}{\partial r} + \hat{\mathbf{e}}_\theta \frac{1}{r}\frac{\partial f}{\partial \theta}. \tag{1.77}$$

1.8.2 Gradient in 3D Cylindrical Coordinates

In cylindrical coordinates (ρ, φ, z) the gradient of a scalar field f is:

$$\nabla f = \mathbf{e}^i \frac{\partial f}{\partial x^i} = \hat{\mathbf{e}}_i \frac{1}{h_i}\frac{\partial f}{\partial x^i} \tag{1.78}$$

$$= \hat{\mathbf{e}}_\rho \frac{\partial f}{\partial \rho} + \hat{\mathbf{e}}_\varphi \frac{1}{\rho}\frac{\partial f}{\partial \varphi} + \hat{\mathbf{e}}_z \frac{\partial f}{\partial z}. \tag{1.79}$$

1.8.3 Gradient in 3D Spherical Coordinates

In spherical coordinates (r, θ, φ) the gradient is:

$$\nabla f = \mathbf{e}^i \frac{\partial f}{\partial x^i} = \hat{\mathbf{e}}_i \frac{1}{h_i}\frac{\partial f}{\partial x^i} \tag{1.80}$$

$$= \hat{\mathbf{e}}_r \frac{\partial f}{\partial r} + \hat{\mathbf{e}}_\theta \frac{1}{r}\frac{\partial f}{\partial \theta} + \hat{\mathbf{e}}_\varphi \frac{1}{r\sin\theta}\frac{\partial f}{\partial \varphi}. \tag{1.81}$$

1.9 Physical Examples

1.9.1 Transformation of a Vector Under Rotation

In physical applications, zero-th order tensors are scalar quantities, such as e.g. the energy of the system, or angles arising in dynamical and kinematic problems. They have the same numerical values in all coordinate frames. A zero-th order tensor can be obtained from two tensors of order one (vectors) by means of a scalar product.

Conversely, first-order tensors can be obtained by zero-th order tensors by acting on the latter with a differential (vector) operator. For example, the electric field (tensor of order one) **E** is obtained from the electrostatic field, a scalar ϕ, by acting on the latter with the gradient ∇ (which is a vector):

$$\mathbf{E} = -\nabla\phi, \tag{1.82}$$

or $E_i = -\frac{\partial\phi}{\partial x_i}$.

Let us consider how the electric field transforms under a rotation of the (Cartesian) coordinate axes in 3D. In particular, we shall consider a rotation about the z-axis by an angle ϑ. From our previous discussion of Euclidean tensors transformation laws we have:

$$E'_i = -\frac{\partial\phi}{\partial x'^i} = -\frac{\partial\phi}{\partial x^j}\frac{\partial x^j}{\partial x'^i} = \frac{\partial x^j}{\partial x'^i}E_j = L_{ij}E_j. \tag{1.83}$$

In the above, we applied the chain rule,

$$\frac{\partial\phi}{\partial x^i} = \frac{\partial\phi}{\partial x'^l}\frac{\partial x'^l}{\partial x^i}, \tag{1.84}$$

which introduces the summation index l. Since l is a dummy summation index, it may be relabeled. Defining

$$L_{ij} = \frac{\partial x'^j}{\partial x^i}, \tag{1.85}$$

the transformation law becomes

$$E_i = L_{ij}E'_j, \tag{1.86}$$

showing explicitly that the electric field transforms as a covariant vector.

Now we implement the transformation of coordinate frames (rotation by an angle ϑ about the z-axis):

$$\begin{cases} x_1 = x'_1\cos\vartheta - x'_2\sin\vartheta \\ x_2 = x'_1\sin\vartheta + x'_2\cos\vartheta \\ x_3 = x'_3 \end{cases} \tag{1.87}$$

to determine the coefficients in the transformation law of the electric field:

$$\frac{\partial x^i}{\partial x'^j} = L_{ij} = \begin{pmatrix} \cos\vartheta & \sin\vartheta & 0 \\ -\sin\vartheta & \cos\vartheta & 0 \\ 0 & 0 & 1 \end{pmatrix}. \tag{1.88}$$

Hence the new components in the rotated frame are given by $E'_i = L_{ij}E_j$ with the coefficients L_{ij} given by Eq. (1.88).

1.9.2 Transformation of a Rank-2 Tensor

Tensors of rank-2 are very common in physics. Well known examples are the strain and stress tensors in elasticity theory, or the electric field gradient. In general, a rank-2 tensor is obtained from the direct (dyadic) product of two rank-1 tensors:

$$T_{ij} = u_i v_j. \tag{1.89}$$

Hence, under a rotation like in the previous example, the new components of the rotated T_{ij} can be easily found by separately rotating u_i and v_j:

$$T'_{ij} = u'_i v'_j = L_{ik}u_k \, L_{jl}v_l = L_{ik}L_{jl}u_k v_l = L_{ik}L_{jl}T_{kl}. \tag{1.90}$$

Therefore, the transformation law in this case reads as:

$$T'_{ij} = L_{ik}L_{jl}T_{kl} \tag{1.91}$$

where the numerical values of the components of L_{ik} and L_{jl} are computed as in the previous example. We should notice how, for a rank-2 tensor, the product of two transformation matrices, L_{ik} and L_{jl} is required, instead of just one for the case of a rank-1 tensor (vector) in the previous example. Hence, for a rank-3 tensor we will need the product of three transformation matrices, and so on.

Another way of writing the direct product Eq. (1.89) is in vector-matrix notation:

$$\mathbf{T} = \mathbf{u} \otimes \mathbf{v} \tag{1.92}$$

with the direct product symbol $\otimes$. By now making the components explicit, we can write:

$$\mathbf{T} = u_i\mathbf{e}_i \otimes v_j\mathbf{e}_j = u_i v_j \mathbf{e}_i \otimes \mathbf{e}_j = T_{ij}\mathbf{e}_i \otimes \mathbf{e}_j. \tag{1.93}$$

Under a rotation of the coordinate frame, like we did above, we thus have, in this notation:

$$\mathbf{T} = T_{ij}\mathbf{e}_i \otimes \mathbf{e}_j = T'_{ij}\mathbf{e}'_i \otimes \mathbf{e}'_j. \tag{1.94}$$

we should emphasize that $\mathbf{T}$, written in coordinate-free notation, represents the same physical object in all reference frames. What changes upon going from one frame to the other, is the numerical value of the components, $T_{ij} \rightarrow T'_{ij}$, which is different in each frame. If the tensor is not a Cartesian tensor, but has components which follow distinct covariant or contravariant transformation laws, using the latter notation we can write:

$$\mathbf{T} = T_{ij}\, \mathbf{e}^i \otimes \mathbf{e}^j \tag{1.95}$$

$$= T^{ij} \mathbf{e}_i \otimes \mathbf{e}_j \tag{1.96}$$

$$= T^i_j\, \mathbf{e}_i \otimes \mathbf{e}^j . \tag{1.97}$$

While this notation is redundant for Euclidean Cartesian tensors, it is, instead, essential for tensors in curved spaces or in flat spaces where the metric has components with different signs, as is the case in Minkowski space-time, which is the topic of the next chapter.

References

1. K. Cahill, *Physical Mathematics*, 2nd edn. (Cambridge University Press, Cambridge, United Kingdom, 2019)
2. K.F. Riley, M.P. Hobson, S.J. Bence, *Mathematical Methods for Physics and Engineering*, 3rd edn. (Cambridge University Press, Cambridge, UK, 2006)

Chapter 2
Minkowski Spacetime

Abstract In this chapter we illustrate the transformation laws introduced in the previous chapter on the example of the Minkowski spacetime of special relativity. Further, we discuss how the tensorial description can be used to formulate the Maxwell's equations in covariant form. A more extensive treatment of the classical theory of fields can be found in Landau and Lifshitz [1].

2.1 Metric

Minkowski space, denoted by $\mathbb{R}^{1,3}$, is a flat four-dimensional space equipped with the metric:

$$[\eta_{kl}] = \left[\eta^{kl}\right] = \mathrm{diag}(-1, 1, 1, 1). \tag{2.1}$$

Minkowski spacetime is the vector space $\mathbb{R}^4$ equipped with the metric tensor $\eta_{\mu\nu}$ of signature $(-, +, +, +)$. From the metric, the scalar product is defined as:

$$(\mathbf{p}, \mathbf{q}) \equiv \mathbf{p} \cdot \mathbf{q} \equiv p^k \eta_{kl} q^l \tag{2.2}$$

where $\mathbf{p} = p^k \mathbf{e}_k$ and $\mathbf{q} = q^l \mathbf{e}_l$. To be consistent with the standard notation for four-vectors, we should write $(p, q) \equiv p^\mu \eta_{\mu\nu} q^\nu$, but for notational simplicity we shall continue to use the vector notation $(\mathbf{p}, \mathbf{q})$, with the understanding that these quantities represent four-vectors in Minkowski spacetime rather than ordinary three-dimensional vectors.

For the metric, we have:

$$\eta_{ij} = (\mathbf{c}_i, \mathbf{c}_j) \tag{2.3}$$

given the index-lowering property Eq. (1.14).

A. Zaccone, *Differential Geometry and Group Theory*,
https://doi.org/10.1007/978-3-032-27518-9_2

From this it follows not only that the basis $\{\mathbf{e}_j\}$ is orthonormal (more precisely, it is pseudo-orthonormal with respect to the Minkowski metric), but also that it is orthonormal to the dual basis:

$$(\mathbf{e}^i, \mathbf{e}_j) = \eta^{ik}(\mathbf{e}_k, \mathbf{e}_j) = \eta^{ik}\eta_{kj} = \delta^i_j. \tag{2.4}$$

For a given vector $\mathbf{x} = x^k\mathbf{e}_k$, if we change to a basis $\{\tilde{\mathbf{e}}_j\}$, the components in the new basis are found as:

$$y^i = \tilde{\mathbf{e}}^i \cdot \mathbf{x} = (\tilde{\mathbf{e}}^i, \mathbf{e}_j)x^j \equiv L^i{}_j x^j \tag{2.5}$$

where we have defined the Lorentz transformation $L^i{}_j \equiv (\tilde{\mathbf{e}}^i, \mathbf{e}_j)$.

The scalar product is invariant under a change of basis, thus:

$$\begin{cases} (\mathbf{p}, \mathbf{q}) = \eta_{ij} p^i q^j \\ (\mathbf{p}, \mathbf{q}) = \eta_{kl}\tilde{p}^k\tilde{q}^l = \eta_{kl} L^k_{\ i} L^l{}_j p^i q^j \end{cases} \iff \eta_{ij} = L^k_{\ i}\eta_{kl}L^l{}_j. \tag{2.6}$$

Writing in a coordinate-free manner ($L^k_{\ i} = (L^\mathsf{T})_i^{\ k}$):

$$\eta = \mathbf{L}^\mathsf{T}\,\eta\,\mathbf{L}. \tag{2.7}$$

Recalling the transformation laws Eqs. (1.6) and (1.7) we have:

$$\begin{aligned} \frac{\partial y^i}{\partial x^j} &= \tilde{\mathbf{e}}^i \cdot \mathbf{e}_j \\ \frac{\partial x^i}{\partial y^j} &= \mathbf{e}^i \cdot \tilde{\mathbf{e}}_j. \end{aligned} \tag{2.8}$$

It is possible to derive a general relationship between the contravariant and covariant transformation laws:

$$\frac{\partial y^i}{\partial x^j} = \tilde{\mathbf{e}}^i \cdot \mathbf{e}_j = \eta^{ik}\eta_{jl}\frac{\partial x^l}{\partial y^k}. \tag{2.9}$$

Since for the Minkowski metric $\eta^{ik}\eta_{jl} = 0, \pm 1$, the transformation coefficients can differ at most by a sign. Since the metric tensor raises and lowers indices, the covariant and contravariant transformation matrices are related through the metric.

It is possible to obtain Lorentz transformations not only as linear maps between coordinates for a change of inertial frame, but also as linear maps between bases; we define:

$$\tilde{\mathbf{e}}_i = \Lambda_i^{\ k}\mathbf{e}_k \tag{2.10}$$

thus:

$$\eta_{ij} = \tilde{\mathbf{e}}_i \cdot \tilde{\mathbf{e}}_j = \Lambda_i^{\ k} \Lambda_j^{\ l} \mathbf{e}_k \cdot \mathbf{e}_l = \Lambda_i^{\ k} \eta_{kl} (\Lambda^\mathsf{T})^l_{\ j} \tag{2.11}$$

or, in coordinate-free form:

$$\eta = \Lambda\, \eta\, \Lambda^\mathsf{T}. \tag{2.12}$$

From the invariance of the scalar product follows the invariance of the modulus, that is:

$$\mathbf{p} = x^k \mathbf{e}_k = y^i \tilde{\mathbf{e}}_i = L^i_{\ k} \Lambda_i^{\ j} x^k \mathbf{e}_j \quad \Longleftrightarrow \quad (\Lambda^\mathsf{T})^j_{\ i} L^i_{\ k} = \delta^j_{\ k} \tag{2.13}$$

or, coordinate-free:

$$\Lambda^\mathsf{T} = \mathbf{L}^{-1}. \tag{2.14}$$

Multiplying Eq. (2.12) on the right and left by η, and using $\eta^2 = \mathbf{I}_4$, we have:

$$\eta\, \Lambda\, \eta\, \Lambda^\mathsf{T} = \Lambda\, \eta\, \Lambda^\mathsf{T}\, \eta. \tag{2.15}$$

From Binet's theorem, we also find:

$$\det \Lambda = \pm 1. \tag{2.16}$$

2.2 Special Relativity

The spacetime of special relativity is described by the Minkowski metric.

Since the separation (distance) between two events in spacetime is invariant, consider two points close to each other $p^i = x^i$ and $q^i = x^i + dx^i$:

$$(\mathbf{p} - \mathbf{q}) \cdot (\mathbf{p} - \mathbf{q}) = dx^i \mathbf{e}_i \cdot dx^j \mathbf{e}_j = \eta_{ij} dx^i dx^j = -(dx^0)^2 + d\mathbf{r}^2. \tag{2.17}$$

Since $dx^0 = c\, dt$, we find the invariance of the line element for timelike worldlines:

$$ds^2 = \eta_{ij} dx^i dx^j = -c^2 dt^2 + d\mathbf{r}^2 = (\mathbf{v}^2 - c^2)\, dt^2 \leq 0. \tag{2.18}$$

With the metric convention $\eta = \text{diag}(-1, 1, 1, 1)$, timelike intervals satisfy $ds^2 < 0$, null intervals satisfy $ds^2 = 0$, and spacelike intervals satisfy $ds^2 > 0$. Equality holds only for massless particles moving at the speed of light.

In the comoving inertial frame $\mathbf{v} = \mathbf{0}$, we can define the proper time τ as:

$$d\tau = \frac{\sqrt{-ds^2}}{c} = \sqrt{1 - \frac{\mathbf{v}^2}{c^2}}\, dt. \tag{2.19}$$

Thus, we deduce time dilation. If a particle has a mean lifetime $T(\mathbf{0})$, in a generic inertial frame moving at relative velocity $\mathbf{v}$, its mean lifetime will be:

$$T(\mathbf{v}) = \gamma T(\mathbf{0}) \tag{2.20}$$

where the Lorentz factor is defined as $\gamma \equiv \left(1 - \frac{\mathbf{v}^2}{c^2}\right)^{-1/2} \in [1, +\infty)$.

2.2.1 Covariant Electrodynamics

Maxwell's equations can be written in manifestly covariant form with only a couple of small adjustments.

The scalar and vector potentials are combined into the four-potential:

$$A_i \equiv \left(-\frac{\phi}{c}, \mathbf{A}\right). \tag{2.21}$$

The minus sign in the temporal component reflects the chosen metric signature $(-, +, +, +)$. In fact, starting from the contravariant four–potential $A^\mu = (\phi/c, \mathbf{A})$ and lowering the index with the Minkowski metric, $A_\mu = \eta_{\mu\nu} A^\nu$, one obtains $A_\mu = (-\phi/c, \mathbf{A})$. The electric field $\mathbf{E} = -\nabla\phi - \dot{\mathbf{A}}$ and the magnetic field $\mathbf{B} = \nabla \times \mathbf{A}$ can be expressed in terms of the four-potential:

$$\begin{aligned} B_i &= \epsilon^{ijk} \partial_j A_k \\ E_i &= c\,(\partial_i A_0 - \partial_0 A_i) \end{aligned} \qquad \text{with} \quad i, j, k = 1, 2, 3. \tag{2.22}$$

It is evident that $\mathbf{E}$ and $\mathbf{B}$ can be combined into a single object, a rank-2 tensor called the Faraday tensor:

$$F_{ij} \equiv \partial_i A_j - \partial_j A_i. \tag{2.23}$$

It is immediately seen that it is an antisymmetric tensor, and that $E_i = cF_{0i}$ and $B^i = \frac{1}{2}\epsilon^{ijk} F_{jk}$, with $i, j, k = 1, 2, 3$. There should be no confusion about the place of index i in the latter expression, because upper/lower placement of indices on the Levi-Civita symbol makes no numerical difference in Cartesian coordinates. Using Proposition 1.3, we find that $F_{jk} = \epsilon_{ijk} B^i$, with $i, j, k = 1, 2, 3$.

Explicitly, given the metric with signature $(-, +, +, +)$, we have:

$$[F_{ij}] = \begin{bmatrix} 0 & -E_1 & -E_2 & -E_3 \\ E_1 & 0 & B_3 & -B_2 \\ E_2 & -B_3 & 0 & B_1 \\ E_3 & B_2 & -B_1 & 0 \end{bmatrix} \tag{2.24}$$

Thanks to the Faraday tensor, the two homogeneous Maxwell equations $\nabla \cdot \mathbf{B} = 0$ and $\nabla \times \mathbf{E} = -\dot{\mathbf{B}}$ can be condensed into a single equation (using the four-dimensional Levi-Civita symbol):

$$\epsilon^{ijkl} \partial_j F_{kl} = 0. \tag{2.25}$$

This is also called the Bianchi identity, and expresses, among other things, the non-existence of magnetic monopoles: in fact, they are topological defects (Dirac monopoles), so their existence would make the Bianchi identity nonzero. This would happen, mathematically, if the Faraday tensor were to pick up a third term on top of the two standard ones of classical electromagnetism. This happens, for example, in certain theories of Dirac monopoles such as the Cabibbo-Ferrari theory where the Faraday tensor reads as:

$$F^{\alpha\beta} = \partial^\alpha A_{\mathrm{e}}^\beta - \partial^\beta A_{\mathrm{e}}^\alpha - \mu_0 c \varepsilon^{\alpha\beta\mu\nu} \partial_\mu A_{\mathrm{m}\nu}, \tag{2.26}$$

where now Greek indices are used to denote the components, and the last term (new with respect to standard classical electromagnetism) is responsible for breaking the Bianchi identity. So far magnetic monopoles have never been observed experimentally apart from analogous emergent quasiparticle excitations observed in certain condensed-matter systems.

In particular, $i = 0$ in Eq. (2.25) gives Gauss's law for $\mathbf{B}$, and $i = 1, 2, 3$ give the spatial components of Faraday's law.

For the inhomogeneous Maxwell equations $\nabla \cdot \mathbf{E} = \frac{\rho}{\epsilon_0}$ and $\nabla \times \mathbf{B} = \mu_0 \mathbf{J} + \frac{1}{c^2} \dot{\mathbf{E}}$, it is necessary to define the four-current:

$$j^k \equiv (c\rho, \mathbf{J}). \tag{2.27}$$

In this way, we can write:

$$\partial_i F^{ki} = \mu_0 j^k. \tag{2.28}$$

The general form of the Lorentz force can be written as:

$$\frac{dp^\mu}{d\tau} = q \left(F^{\mu 0} + F^{\mu j} v_j \right). \tag{2.29}$$

It is also possible to write the Lorentz force only for spatial components:

$$\frac{dp^i}{dt} = q\left(-F^{i0} + \epsilon^{ijk} v_j B_k\right), \tag{2.30}$$

where $i = 1, 2, 3$. Selecting, instead, $\mu = 0$, and recalling that $p^0 = \frac{\mathcal{E}}{c}$, with $\mathcal{E}$ being energy, we have $\dot{\mathcal{E}} = q\,\mathbf{E} \cdot \mathbf{v}$, confirming the fact that the magnetic field $\mathbf{B}$ does no work.

Writing Maxwell's equations in tensorial form ensures that Einstein's first postulate is satisfied, namely that the laws of physics are the same in any inertial reference frame, since they are invariant under Lorentz transformations.

For electrodynamics to be a fully relativistic theory, two corrections must be made:

1. The momentum considered must be the relativistic momentum $\mathbf{p} = \gamma m\mathbf{v}$;
2. The total energy must include the rest energy $\mathcal{E}_0 = mc^2$.

Apart from these two corrections, however, the classical electrodynamics described by the Maxwell equations is already in a "naturally invariant" form.

2.3 Physical Examples

2.3.1 Gauss's Law for Magnetism and Faraday's Law from Eq. (2.25)

Show that Eq. (2.25) indeed recovers the two homogeneous Maxwell equations.

First of all, it is convenient to rewrite the equation using Greek indices and use Latin indices for the spatial components only:

$$\epsilon^{\alpha\lambda\mu\nu}\,\partial_\lambda F_{\mu\nu} = 0, \tag{2.31}$$

where each Greek index runs over $0, 1, 2, 3$.

Next we separate the Faraday tensor into its temporal (0) and spatial (1,2,3) components, the latter denoted by Latin indices. This gives us the electric and magnetic fields as:

$$F_{00} = 0, \qquad F_{k0} = -F_{0k} = E_k, \qquad F_{ij} = \epsilon_{ijk} B^k. \tag{2.32}$$

If we set $\alpha = 0$ we get

$$\epsilon^{0ijk}\,\partial_i F_{jk} = \epsilon^{ijk}\,\partial_i(\epsilon_{jkp} B^p) = 2\,\partial_i B^i = 2\,\nabla\cdot\mathbf{B} = 0, \tag{2.33}$$

because $\epsilon^{ijk}\epsilon_{jkp} = 2\,\delta^i{}_p$. Hence, we have obtained Gauss's law for magnetism.

For $\alpha = i$ we have

$$\epsilon^{ij0k}\,\partial_j F_{0k} + \epsilon^{ijk0}\,\partial_j F_{k0} + \epsilon^{i0jk}\,\partial_0 F_{jk} = 0,$$

$$-\,2\,\epsilon^{ijk}\,\partial_j E_k - \epsilon^{ijk}\,\partial_0(\epsilon_{jkp}B^p) = 0,$$

and therefore

$$\epsilon^{ijk}\,\partial_j E_k + \partial_0 B^i = 0, \tag{2.34}$$

which represents Faraday's law of induction.

2.3.2 Gauss and Ampere Laws Under a Change of Coordinate Frame

We can recall the Gauss-Ampere law Eq. (2.28) now written for the covariant form of the 4-current vector j_k:

$$\partial_i F_{ki} = \mu_0 j_k \tag{2.35}$$

with components measured in a Lorentzian frame $\mathcal{K}$. We now move to a new Lorentzian frame $\mathcal{K}'$. Because physical laws are invariant, the form of the equation will be exactly the same and only the numerical components of the tensorial quantities involved will change:

$$\partial_l F'_{ml} = \mu_0 j'_m. \tag{2.36}$$

The numerical values of the components change according to the transformation law for covariant vectors in Minkowski space, which for a generic covariant vector A_i read as:

$$A'_i = (L^j_i)^{-1} A_j = \Lambda^j_i A_j \tag{2.37}$$

with $\Lambda^j_i = \partial x_j/\partial x'_i$. Hence, in terms of numerical quantities defined in the original frame $\mathcal{K}$:

$$\Lambda^k_m\,(\partial_i F_{ki} - \mu_0 j_k) = 0, \tag{2.38}$$

which transforms like a covariant vector of components $m = 0, 1, 2, 3$. We thus see that the form of the Gauss-Ampere law Eq. (2.7) remains the same in all Lorentz frames, and this is ultimately ensured by the fact that both the left hand side and the right hand side of the equation transform in the same way, i.e. they are both rank-1 covariant tensors. Hence, it is ultimately the tensorial form of the equation, when properly balanced, which makes it valid in all coordinate frames that share the same properties (i.e. Lorentzian frames, in our examples). In the physics jargon, it is often said physical laws, such as the Maxwell laws, are in *covariant* form, although this terminology does not refer to the specific covariant transformation law but simply indicates that laws are invariant with respect to the choice of reference frame.

Reference

1. L.D. Landau, E.M. Lifshitz, *The Classical Theory of Fields*, Course of Theoretical Physics, vol. 2, 4th edn. (Butterworth-Heinemann, Oxford, 1975)

Chapter 3
Differential Forms

Abstract In this chapter we introduce the language of differential forms. This is a set of tools which will allow us to produce useful generalizations of geometric relations, such as, for example, high-dimensional hypervolume elements and differential geometry relations for differential operators in curved spaces. We begin by introducing the wedge product, which allows us to construct antisymmetric tensors representing oriented geometric objects such as areas and volumes. We then define the exterior derivative, which generalizes the familiar gradient, curl, and divergence operators. Finally, we introduce the Hodge dual, which provides a metric–dependent duality between forms and allows many physical equations, such as Maxwell's equations, to be written in a compact coordinate–independent form. For more details, the interested reader is referred to Szekeres [1] and Cahill [2].

3.1 Wedge Product

Differential forms provide a coordinate-independent language for integration, orientation, and differential operators. They allow geometric quantities such as line elements, surface elements, and volume elements to be treated in a unified way, independent of the dimension of the underlying space.

In this framework, vectors and tensors are replaced by objects that are naturally integrated over curves, surfaces, and higher-dimensional manifolds. We begin by introducing the simplest of these objects: 1-forms.

Definition 3.1 A *1-form* is a covector field, i.e. a rank-1 covariant tensor, which in local coordinates can be written as

$$\omega_1 \equiv a_k \, dx^k. \tag{3.1}$$

A. Zaccone, *Differential Geometry and Group Theory*,
https://doi.org/10.1007/978-3-032-27518-9_3

1-forms are invariant under changes of reference frame:

$$\tilde{\omega}_1 = \tilde{a}_k dy^k = \frac{\partial x^i}{\partial y^k} a_i \frac{\partial y^k}{\partial x^j} dx^j = \delta^i_j a_i dx^j = a_i dx^i = \omega_1 . \tag{3.2}$$

It is also possible to define 0-forms, but they trivially coincide with scalar fields.

To describe oriented geometric objects such as areas and volumes, we need a product that keeps track of orientation. This leads naturally to an antisymmetric product between differentials, known as the wedge product.

One can define an antisymmetric product between forms, the wedge product:

$$\begin{aligned} dx \wedge dy &= -dy \wedge dx \\ dx \wedge dx &= dy \wedge dy = 0. \end{aligned} \tag{3.3}$$

Geometrically, the wedge product represents oriented area elements. For instance, the expression $dx \wedge dy$ can be interpreted as the infinitesimal oriented area spanned by the coordinate directions x and y. The antisymmetry property ensures that reversing the order of the coordinates reverses the orientation of the area element.

Antisymmetry is necessary in order for coordinates to transform correctly under changes of reference frame.

Geometrically, the wedge product is associated with an oriented area; for example, in two dimensions with a change of reference frame $(x, y) \rightarrow (u, v)$:

$$du \wedge dv = \left(\frac{\partial u}{\partial x}dx + \frac{\partial u}{\partial y}dy\right) \wedge \left(\frac{\partial v}{\partial x}dx + \frac{\partial v}{\partial y}dy\right) = \underbrace{\left(\frac{\partial u}{\partial x}\frac{\partial v}{\partial y} - \frac{\partial v}{\partial x}\frac{\partial u}{\partial y}\right)}_{\det \mathsf{J}} dx \wedge dy \tag{3.4}$$

which is precisely the transformation law for surface areas that we know from elementary calculus, involving the Jacobian matrix J of the coordinate transformation.

Generalizing to n dimensions:

$$dy_1 \wedge \cdots \wedge dy_n = (\det \mathsf{J})\, dx_1 \wedge \cdots \wedge dx_n . \tag{3.5}$$

This relation is valid in general and not only for orthogonal systems (in which case wedge products would be replaced by standard product).

Definition 3.2 A 2-*form* is defined as the contraction of a rank-2 tensor with a double wedge product:

$$\omega_2 \equiv \frac{1}{2} a_{ij} dx^i \wedge dx^j . \tag{3.6}$$

Definition 3.3 A *p-form* is defined as the contraction of a rank-p tensor:

$$\omega_p \equiv \frac{1}{p!} a_{i_1 \ldots i_p} dx^{i_1} \wedge \cdots \wedge dx^{i_p} \tag{3.7}$$

Proposition 3.1 *The number of independent components of a p-form on an n-dimensional space is* $\binom{n}{p}$.

This follows from the antisymmetry of the wedge product, which can be expressed in terms of tensor products:

$$dx^i \wedge dx^j \equiv dx^i \otimes dx^j - dx^j \otimes dx^i . \tag{3.8}$$

In compact form, one can write:

$$dx^{i_1} \wedge \cdots \wedge dx^{i_p} \equiv dx^{[i_1} \wedge \cdots \wedge dx^{i_p]} \tag{3.9}$$

where square brackets denote antisymmetrization of the indices (sum over even permutations and subtraction over odd ones). From this, the factor $\frac{1}{p!}$ in Definition 3.3 is explained: starting from a tensor with no symmetries, the construction of a p-form through the wedge product extracts only its antisymmetric part.

3.2 Exterior Derivative

A differential operator between forms can be defined, the exterior derivative d, which is a coordinate-independent operator generalizing the gradient, curl, and divergence:

$$d\omega_p = \frac{1}{p!} \partial_k a_{i_1 \ldots i_p} dx^k \wedge dx^{i_1} \wedge \cdots \wedge dx^{i_p} . \tag{3.10}$$

Thus, the exterior derivative transforms a p-form into a $(p+1)$-form.

3.2.1 The Exterior Derivative Anticommutes with the Wedge Product

It can also be shown that the exterior derivative anticommutes with the wedge product.

Proposition 3.2 *Given a p-form* ω_p *and a generic form* ς, *one has:*

$$d(\omega_p \wedge \varsigma) = d\omega_p \wedge \varsigma + (-1)^p \omega_p \wedge d\varsigma \tag{3.11}$$

Proof In the case of two 1-forms $\omega = a_i dx^i$ and $\varsigma = b_j dx^j$:

$$\begin{aligned}
d(\omega \wedge \varsigma) &= d\left(a_i dx^i \wedge b_j dx^j\right) = \partial_k (a_i b_j) dx^k \wedge dx^i \wedge dx^j \\
&= (\partial_k a_i) b_j dx^k \wedge dx^i \wedge dx^j + a_i (\partial_k b_j) dx^k \wedge dx^i \wedge dx^j \\
&= (\partial_k a_i) b_j dx^k \wedge dx^i \wedge dx^j - a_i (\partial_k b_j) dx^i \wedge dx^k \wedge dx^j \\
&= \left(\partial_k a_i dx^k \wedge dx^i\right) \wedge b_j dx^j - a_i dx^i \wedge \left(\partial_k b_j dx^k \wedge dx^j\right) \\
&= d\omega \wedge \varsigma - \omega \wedge d\varsigma.
\end{aligned}$$

□

3.2.2 Example

Given a scalar field (0-form) f, its differential is $df = \frac{\partial f}{\partial x^k} dx^k$.

Given a 1-form on $\mathbb{R}^3$: $A = A_x dx + A_y dy + A_z dz$, one finds that its exterior derivative is $dA = (\nabla \times \mathbf{A})_x \, dy \wedge dz + (\nabla \times \mathbf{A})_y \, dz \wedge dx + (\nabla \times \mathbf{A})_z \, dx \wedge dy$, where $\mathbf{A} = \left(A_x, A_y, A_z\right)$ has been defined.

3.2.3 Example

For a 2-form on $\mathbb{R}^2$: $\omega_2 = \frac{1}{2} A_{ij} dx^i \wedge dx^j = \frac{1}{2} (A_{12} - A_{21}) \, dx \wedge dy$. Thus, if A_{ij} is antisymmetric this reduces to $\omega_2 = A_{12} dx \wedge dy$, whereas if it is symmetric one has $\omega_2 = 0$.

3.3 Bianchi Identity, Exact and Closed Forms

One of the most important properties of the exterior derivative is that applying it twice always yields zero. This identity encodes deep geometric and physical constraints and is known as the Bianchi identity.

Proposition 3.3 *Under suitable regularity assumptions, one has $dd(\ldots) \equiv 0$ (Bianchi identity).*

Proof Given $\omega_{p+1} = d\omega_p$:

$$dd\,\omega_p = d\omega_{p+1} = \left(\partial_j \partial_k a_{i_1 \ldots i_p}\right) dx^j \wedge dx^k \wedge dx^{i_1} \wedge \cdots \wedge dx^{i_p}$$

By Schwarz's lemma $\partial_j \partial_k f$ is symmetric for regular f, hence by Eq. (3.8) one has $dd\,\omega_p = 0$. □

The relation $dd(\ldots) = 0$ is an elegant way of writing the Bianchi identity: in order to break this identity, a perturbation of the underlying space is required, known as a topological defect (cf. the discussion about Dirac monopoles and the Cabibbo-Ferrari theory in Chap. 2).

The property $dd(\ldots) = 0$ allows us to classify differential forms based on their differential structure. Accordingly, one can further characterize differential forms:

Definition 3.4 A p-form ω_p is said to be closed if $d\omega_p = 0$.

Definition 3.5 A p-form ω_p is said to be exact if there exists a $(p-1)$-form ω_{p-1} such that $\omega_p = d\omega_{p-1}$.

Proposition 3.4 *Every exact form is also closed.*

Proof Directly from Proposition 3.3. □

Lemma 3.1 (Poincaré) *On a simply connected manifold, every closed form is also exact.*

Theorem 3.1 (Stokes) *Given a p-form ω and a simply connected domain S, one has:*

$$\int_{\partial S} \omega = \int_S d\omega \tag{3.12}$$

This theorem unifies the fundamental integral theorems of vector calculus (Green, Gauss, and Stokes) into a single geometric statement.

3.3.1 Physical Examples

3.3.1.1 Static Electric Fields

A classical electrostatic field is closed and also exact. Because $\dot{\mathbf{B}} = 0$, then by Faraday's law $\nabla \times \mathbf{E} = 0$. We can write the electric field as a 1-form,

$$E = E_i\, dx^i, \qquad i = 1, 2, 3 \tag{3.13}$$

$$\Rightarrow \quad dE = \partial_j E_i\, dx^j \wedge dx^i \tag{3.14}$$

$$= \tfrac{1}{2}\big(\partial_j E_i - \partial_i E_j\big) dx^j \wedge dx^i \tag{3.15}$$

$$= 0 \tag{3.16}$$

hence the equality $dE = 0$ expresses the fact that the curl of the electric field is zero and, also, that the electric field is *closed*. We can now define the line integral of the 1-form E along a pathway P:

$$V_P(\mathbf{x}) = -\int_{P,\mathbf{x}_0}^{\mathbf{x}} E_i\, dx^i = -\int_P E. \tag{3.17}$$

However, because the electric field is irrotational (by assumption), $\nabla \times \mathbf{E} = 0$, then this integral cannot depend on the chosen path, because if we were to follow a different path P' instead of P, then by Stokes' theorem we would have:

$$V_{P'}(\mathbf{x}) - V_P(\mathbf{x}) = \oint_{P' \to P} E_i\, dx^i = \iint_S (\nabla \times \mathbf{E}) \cdot d\mathbf{a} = 0 \tag{3.18}$$

where $d\mathbf{a}$ denotes the surface element. Using Stokes' theorem written in the language of differential forms introduced above (Theorem 3.1):

$$V_{P'}(\mathbf{x}) - V_P(\mathbf{x}) = \int_{\partial S} E = \int_S dE. \tag{3.19}$$

Hence, the electrostatic potential $V_P(\mathbf{x}) \equiv V(\mathbf{x})$ is independent of the integration pathway P such that $\mathbf{E} = -\nabla V(\mathbf{x})$ and the 1-form:

$$E = E_i dx^i = -\partial_i V dx^i = -dV \tag{3.20}$$

is an exact form.

In this example we used the version of Eq. (3.12) appropriate for a 1-form, hence for $p = 1$ (vector fields). The usefulness of the language of differential form lies in its ability to directly generalize to higher-order forms and space dimensions. In this example, the generalized Stokes theorem (Theorem 3.1) is valid for any generic p-form where p is any positive integer. This clearly becomes very handy when one is working in a high-dimensional space (e.g. Kaluza-Klein theories and their modern versions), where a derivation of Stokes' theorem from intuitive geometry as we know it from standard calculus, is impossible.

3.3.1.2 Covariant Electrodynamics

In the language of differential forms, the vector potential of classical electrodynamics is associated to a 1-form: $A = A_j dx^j$. Acting with the exterior derivative onto A, one obtains:

$$dA = dA_j \wedge dx^j = \partial_i A_j\, dx^i \wedge dx^j = \tfrac{1}{2} F_{ij}\, dx^i \wedge dx^j = F \tag{3.21}$$

where we used the short-hand notation $\partial_i \equiv \partial/\partial x_i$. Hence, the exterior derivative transforms the electromagnetic vector potential (or 4-vector gauge field), which is a 1-form, into the Faraday tensor, which is a 2-form.

In the previous chapter we derived Maxwell's equations of electromagnetism in manifestly covariant form:

$$\begin{cases} \epsilon^{\ell ijk}\, \partial_i F_{jk} = 0 \\ \partial_i F_{ki} = \mu_0 j_k \end{cases} \tag{3.22}$$

which condense the four Maxwell equations into just two equations relating the derivatives of the Faraday tensor $\mathbf{F}$ to the current density $\mathbf{j}$. The first equation, which condenses the two homogeneous Maxwell equations (Gauss's law for magnetism and Faraday's law), can be compactly written using the language of differential forms as just:

$$dF = ddA = 0. \tag{3.23}$$

This formally represents a Bianchi identity, $dd(\ldots) = 0$, and stipulates that F is an exact form, and, therefore (Proposition 3.3), is also closed ($dF = 0$).

3.3.1.3 Example

Justify, with all the algebraic steps, the presence of a factor 1/2 after the third equality in Eq. (3.21), and hence in the definition of *p-form* in (3.3).

Start with

$$F = \sum_{i,j} \partial_i A_j\, dx^i \wedge dx^j, \tag{3.24}$$

where the double sum over the repeated indices i and j is spelled out explicitly.

Now we split the double sum as

$$F = \sum_{i<j} \partial_i A_j\, dx^i \wedge dx^j + \sum_{i>j} \partial_i A_j,\, dx^i \wedge dx^j + \sum_{i=j} \partial_i A_i,\, dx^i \wedge dx^i. \tag{3.25}$$

But $dx^i \wedge dx^i = 0$, so the last sum vanishes.

Now let us rename indices in the $i > j$ sum by swapping $i \leftrightarrow j$ (just dummy indices):

$$\sum_{i>j} \partial_i A_j,\, dx^i \wedge dx^j = \sum_{i<j} \partial_j A_i,\, dx^j \wedge dx^i. \tag{3.26}$$

Then we use antisymmetry of the wedge product, $dx^j \wedge dx^i = -dx^i \wedge dx^j$:

$$F = \sum_{i<j} \left(\partial_i A_j, dx^i \wedge dx^j - \partial_j A_i, dx^i \wedge dx^j \right) \tag{3.27}$$

$$= \sum_{i<j} \left(\partial_i A_j - \partial_j A_i \right), dx^i \wedge dx^j. \tag{3.28}$$

This already shows that only the antisymmetric part contributes.

Each unordered pair $\{i, j\}$ with $i \neq j$ appears twice in $\sum_{i,j}$ (once as (i, j), once as (j, i)). So

$$\sum_{i<j} (\cdots) = \frac{1}{2} \sum_{i \neq j} (\cdots) = \frac{1}{2} \sum_{i,j} (\cdots), \tag{3.29}$$

because the $(\cdots)$ is antisymmetric and vanishes when $i = j$.

Hence

$$\sum_{i,j} \partial_i A_j, dx^i \wedge dx^j = \frac{1}{2} \sum_{i,j} \left(\partial_i A_j - \partial_j A_i \right), dx^i \wedge dx^j \tag{3.30}$$

$$= \frac{1}{2}, F_{ij}, dx^i \wedge dx^j, \tag{3.31}$$

with $F_{ij} = \partial_i A_j - \partial_j A_i$.

So the factor $1/2$ is exactly the correction for double counting the pairs (i, j) and (j, i) in the full double sum due to the basis $dx^i \wedge dx^j$ being antisymmetric. The same demonstration can be extended to a generic *p-form* simply by iteration.

3.4 Hodge Operator

Up to this point, the algebra of differential forms has been defined independently of any metric structure. The wedge product and exterior derivative depend only on the smooth structure of the manifold. However, many physical applications require the introduction of a metric, which allows one to define notions such as orthogonality and volume. Once a metric is specified, it becomes possible to relate p-forms to $(n-p)$-forms through a natural duality operator known as the Hodge star. While the wedge product allows us to build higher-dimensional geometric objects, it does not by itself relate forms of different degrees. To construct such a relation, additional structure is required: namely, a metric and an orientation. This leads to the definition of the Hodge operator, which implements a natural duality between forms of complementary degree.

3.4.1 Levi-Civita Tensor

The three-dimensional Levi-Civita symbol (Definition 1.12) is totally antisymmetric and isotropic (Eq. (1.12)), and transforms as a tensor density of weight $w = -1$ as shown in Chap. 1. It can be turned into a true rank-3 pseudotensor of weight $w = 0$, by means of multiplication with the square root of the determinant of the metric tensor. This is because the square root of the determinant of the metric tensor transforms as a tensor density of weight $w = +1$, i.e. with an extra factor $|\det\mathbf{R}|^{-1}$. Hence, by means of the multiplication with $\sqrt{g}$, we obtain a genuine tensor tied to the metric, that can be used to express a coordinate-invariant volume element, as we shall see later on.

Definition 3.6 The *Levi-Civita tensor* is defined as $\varepsilon_{ijk} \equiv \sqrt{g}\,\epsilon_{ijk}$.

To find the transformation law of the Levi-Civita tensor, some results from linear algebra are required.

Theorem 3.2 (Laplace) *Given* $\mathsf{M} \in \mathbb{R}^{n\times n}$, *one has* $\det\mathsf{M} = \epsilon^{i_1\ldots i_n} M_{i_1 1}\ldots M_{i_n n}$.

Corollary 3.1 $\epsilon_{k_1\ldots k_n}\det\mathsf{M} = \epsilon^{i_1\ldots i_n} M_{i_1 k_1}\ldots M_{i_n k_n}$.

Corollary 3.2 $\det(\mathsf{M}_1\ldots\mathsf{M}_k) = (\det\mathsf{M}_1)\ldots(\det\mathsf{M}_k)$.

Considering a change of reference frame $x^i \mapsto y^i$:

$$
\begin{aligned}
\varepsilon'_{ijk} &= \underbrace{\sqrt{g'}\,\epsilon'_{ijk} \;=\; \sqrt{g'}\,\epsilon_{ijk}}_{\text{isotropy of }\epsilon_{ijk}} \\
&= \sqrt{\left|\det\left(\frac{\partial x^s}{\partial y^m}\frac{\partial x^t}{\partial y^n}g_{st}\right)\right|}\,\epsilon_{ijk} \\
&= \sqrt{\left|\det\left(\frac{\partial x^s}{\partial y^m}\right)\det\left(\frac{\partial x^t}{\partial y^n}\right)\det(g_{st})\right|}\,\epsilon_{ijk} \qquad (3.32)\\
&= \left|\det\left(\frac{\partial x}{\partial y}\right)\right|\sqrt{g}\,\epsilon_{ijk} \;=\; \sigma\sqrt{g}\,\underbrace{\det\left(\frac{\partial x}{\partial y}\right)\epsilon_{ijk}}_{\text{Corollary 3.1}} \\
&= \sigma\sqrt{g}\,\frac{\partial x^s}{\partial y^i}\frac{\partial x^t}{\partial y^j}\frac{\partial x^u}{\partial y^k}\epsilon_{stu} \;=\; \sigma\,\frac{\partial x^s}{\partial y^i}\frac{\partial x^t}{\partial y^j}\frac{\partial x^u}{\partial y^k}\varepsilon_{stu}.
\end{aligned}
$$

Since ε_{ijk} is defined from ϵ_{ijk} (i.e. from an antisymmetric product), it is a pseudotensor, and therefore its transformation law depends on σ, the sign of the determinant of the Jacobian. Recall that $\det\mathsf{J}$ changes sign for each inversion in the change of reference frame.

The contravariant components of the Levi-Civita tensor are:

$$\begin{aligned}\varepsilon^{ijk} &= g^{is} g^{jt} g^{ku} \varepsilon_{stu} = g^{is} g^{jt} g^{ku} \sqrt{g}\, \epsilon_{stu} = \sqrt{g} \det\left(g^{mn}\right) \epsilon_{ijk} \\ &= \frac{\sqrt{g}}{\det\left(g_{mn}\right)} \epsilon_{ijk} = \frac{\sigma}{\sqrt{g}} \epsilon_{ijk} \end{aligned} \tag{3.33}$$

where $g \equiv |\det(g_{mn})|$ denotes the absolute value of the determinant of the metric tensor.

All these properties of ε_{ijk} also hold for the general n-dimensional Levi-Civita tensor, analogously defined as:

$$\varepsilon_{i_1 \ldots i_n} \equiv \sqrt{g}\, \epsilon_{i_1 \ldots i_n} . \tag{3.34}$$

3.4.2 *Hodge Star*

The Hodge operator, also called the Hodge dual or Hodge star, is an operator that maps a p-form into an $(n-p)$-form, where n is the dimension of the manifold considered.

Definition 3.7 Given an n-dimensional differential manifold $\mathcal{M}$, the *Hodge operator* is defined as $* : \bigwedge^{p} \mathcal{M} \to \bigwedge^{n-p} \mathcal{M}$ such that:

$$*\left(dx^{i_1} \wedge \cdots \wedge dx^{i_p}\right) \equiv \frac{1}{(n-p)!} g^{i_1 k_1} \ldots g^{i_p k_p} \varepsilon_{k_1 \ldots k_p m_1 \ldots m_{n-p}} dx^{m_1} \wedge \cdots \wedge dx^{m_{n-p}} . \tag{3.35}$$

Geometrically, the Hodge operator maps a p-form to the unique $(n-p)$-form that represents its orthogonal complement with respect to the metric-induced volume element.

For a p-form ω, this also implies the master identity:

$$\omega \wedge *\omega = \langle \omega, \omega \rangle\, dV \tag{3.36}$$

where the inner product $\langle , \rangle$ is induced by the metric g.

3.5 Examples

3.5.1 *3D Euclidean Space*

On $\mathbb{R}^3$ one has $*1 = dx \wedge dy \wedge dz$, $*(dx \wedge dy \wedge dz) = 1$, $*dx = dy \wedge dz$, $*(dx \wedge dy) = dz$, and so on. Moreover, two important properties are found:

$$*df = *\left(\frac{\partial f}{\partial x}dx + \frac{\partial f}{\partial y}dy + \frac{\partial f}{\partial z}dz\right) = \frac{\partial f}{\partial x}dy \wedge dz + \frac{\partial f}{\partial y}dz \wedge dx + \frac{\partial f}{\partial z}dx \wedge dy$$

$$*\,\mathbf{F} = *\left(F_x dx + F_y dy + F_z dz\right) = F_x dy \wedge dz + F_y dz \wedge dx + F_z dx \wedge dy$$

Hence one sees that:

$$d * df = \nabla^2 f \, dx \wedge dy \wedge dz \tag{3.37}$$

$$d * \mathbf{F} = \nabla \cdot \mathbf{F} \, dx \wedge dy \wedge dz. \tag{3.38}$$

3.5.2 Minkowski Space

In $\mathbb{R}^{1,3}$ (Minkowski space-time) one must take into account the negative sign of the time coordinate; for instance $*1 = dt \wedge dx \wedge dy \wedge dz$, but $*\,(dt \wedge dx \wedge dy \wedge dz) = -1$. Furthermore, we work in 4D Minkowski space with signature $(-, +, +, +)$, orientation $dt \wedge dx \wedge dy \wedge dz > 0$, and volume element $dV = dt \wedge dx \wedge dy \wedge dz$.

Moreover, for 1-forms:

$$*\,dt = -dx \wedge dy \wedge dz \qquad *dx = -dy \wedge dz \wedge dt$$

$$*dy = -dz \wedge dx \wedge dt \qquad *dz = -dx \wedge dy \wedge dt$$

For 2-forms:

$$*\,(dx \wedge dt) = dy \wedge dz \qquad *(dx \wedge dy) = -dz \wedge dt$$

$$*(dy \wedge dt) = dz \wedge dx \qquad *(dy \wedge dz) = -dx \wedge dt$$

$$*(dz \wedge dt) = dx \wedge dy \qquad *(dz \wedge dx) = -dy \wedge dt$$

For 3-forms:

$$*\,(dx \wedge dy \wedge dz) = -dt \qquad *(dy \wedge dz \wedge dt) = -dx$$

$$*(dz \wedge dx \wedge dt) = -dy \qquad *(dx \wedge dy \wedge dt) = -dz.$$

The following example illustrates how the metric signature and the chosen orientation jointly determine the signs appearing in Hodge duals. Indeed, in the above relations, one may wonder about the sign e.g. why $*(dt \wedge dx) = -\,dy \wedge dz$ whereas $*(dt \wedge dy) = \ dx \wedge dz$. Let us provide a detailed justification for this result, which illustrates the general procedure to correctly compute the sign in the above relations.

Recalling that the inner product is induced by the metric

$$\langle dt \wedge dx, dt \wedge dx\rangle = g^{00}g^{11} = -1, \qquad \langle dt \wedge dy, dt \wedge dy\rangle = g^{00}g^{22} = -1.$$

(1) Compute $*(dt \wedge dx)$. Assume $*(dt \wedge dx) = s\,dy \wedge dz$ with $s = \pm 1$. Then

$$(dt \wedge dx) \wedge *(dt \wedge dx) = s\,dt \wedge dx \wedge dy \wedge dz = s\,dV.$$

By the defining identity this must equal $(-1)\,dV$, hence $s = -1$ and

$$*\,(dt \wedge dx) = -\,dy \wedge dz\,.$$

(2) Compute $*(dt \wedge dy)$. Assume $*(dt \wedge dy) = s\,dx \wedge dz$ with $s = \pm 1$. Then

$$\begin{aligned}(dt \wedge dy) \wedge *(dt \wedge dy) &= s\,dt \wedge dy \wedge dx \wedge dz\\ &= -s\,dt \wedge dx \wedge dy \wedge dz \quad \text{(swap } dy \text{ past } dx\text{)}\\ &= -s\,dV.\end{aligned}$$

Again this must equal $(-1)\,dV$, so $-s = -1$ and $s = +1$, hence

$$*\,(dt \wedge dy) = \ dx \wedge dz\,.$$

Hence, we can summarize the set of relations which express the action of the Hodge operator in Minkowski space as follows:

$$\begin{aligned}
*1 &= \frac{1}{4!}\,\eta_{k\lambda\mu\nu}\,dx^k \wedge dx^\lambda \wedge dx^\mu \wedge dx^\nu,\\
*dx^i &= \frac{1}{3!}\,g^{ik}\,\eta_{k\lambda\mu\nu}\,dx^\lambda \wedge dx^\mu \wedge dx^\nu,\\
*(dx^i \wedge dx^j) &= \frac{1}{2!}\,g^{ik}g^{j\lambda}\,\eta_{k\lambda\mu\nu}\,dx^\mu \wedge dx^\nu,\\
*(dx^i \wedge dx^j \wedge dx^k) &= g^{it}g^{ju}g^{kv}\,\eta_{tuvw}\,dx^w,\\
*(dx^i \wedge dx^j \wedge dx^k \wedge dx^\ell) &= g^{it}g^{ju}g^{kv}g^{\ell w}\,\eta_{tuvw} \;=\; \eta^{ijk\ell}.
\end{aligned}$$

The formalism of differential forms provides an elegant framework for classical field theories. In particular, Maxwell's equations acquire a remarkably compact and geometrically transparent form when expressed in terms of exterior derivatives and Hodge duals.

3.6 Physical Example: Covariant Electrodynamics with Differential Forms

Recalling the gauge vector potential (2.21), it is possible to define the 1-form $A = A_i dx^i$ such that acting with the exterior derivative onto it, we get:

$$dA = \partial_i A_j \, dx^i \wedge dx^j = \frac{1}{2} F_{ij} \, dx^i \wedge dx^j \equiv F \tag{3.39}$$

where the 2-form determined by the Faraday tensor, Eq. (2.23), has been previously defined. Equation (2.25) then becomes, in view of the Bianchi identity:

$$dF = 0. \tag{3.40}$$

In the static case, $\dot{\mathbf{B}} = 0$, so that $\nabla \times \mathbf{E} = 0$, i.e. $dE = 0$: the electrostatic field is closed and, by Poincaré's lemma, locally exact.

It is also possible to compute the Hodge dual of F:

$$*F = \frac{1}{2} F_{ij} *\left(dx^i \wedge dx^j\right) = \frac{1}{4} F_{ij} g^{ik} g^{jl} \varepsilon_{klmn} \, dx^m \wedge dx^n = \frac{1}{4} F^{kl} \varepsilon_{klmn} \, dx^m \wedge dx^n. \tag{3.41}$$

Applying the Hodge dual of the exterior derivative:

$$\begin{aligned}
* \, d * F &= \frac{1}{4} \, \partial_p \left(F^{kl} \varepsilon_{klmn}\right) * \left(dx^p \wedge dx^m \wedge dx^n\right) \\
&= \frac{1}{4} \, \partial_p \left(\sqrt{g} \, F^{kl}\right) \epsilon_{klmn} g^{p\alpha} g^{m\beta} g^{n\gamma} \epsilon_{\alpha\beta\gamma\delta} \, dx^\delta \\
&= \frac{1}{4} \, \partial_p \left(\sqrt{g} \, F^{kl}\right) \sqrt{g} \, \epsilon_{klmn} g^{\alpha p} g^{\beta m} g^{\gamma n} g^{\delta w} \epsilon_{\alpha\beta\gamma\delta} \, dx_w \\
&= \frac{1}{4} \, \partial_p \left(\sqrt{g} \, F^{kl}\right) \sqrt{g} \, \epsilon_{klmn} \frac{\upsilon}{g} \, \epsilon^{pmnw} \, dx_w \\
&= \frac{1}{4} \, \partial_p \left(\sqrt{g} \, F^{kl}\right) \frac{\sigma}{\sqrt{g}} \, 2 \left(\delta^p_k \delta^w_l - \delta^w_k \delta^p_l\right) dx_w \\
&= \frac{\sigma}{2\sqrt{g}} \left[\partial_p \left(\sqrt{g} \, F^{pw}\right) dx_w - \partial_p \left(\sqrt{g} \, F^{wp}\right) dx_w\right] \\
&= -\frac{\sigma}{\sqrt{g}} \, \partial_p \left(\sqrt{g} \, F^{kp}\right) dx_k
\end{aligned} \tag{3.42}$$

where at some point we used the Einstein identity: $\epsilon_{klmn} \, \epsilon^{pmnw} = \left(\delta^p_k \delta^w_l - \delta^w_k \delta^p_l\right)$.

In Minkowski spacetime, $g_{ij} = \mathrm{diag}(-1, 1, 1, 1)$, Eq. (3.42) reduces to

$$* \, d * F = \partial_p (F^{kp}) \, dx_k.$$

Recalling from Eq. (2.28) that $\partial_p(F^{kp}) = \mu_0 j^k$, and defining the current 1-form $j \equiv j^k dx_k$, we obtain the generalization on a curved manifold of the second covariant Maxwell equation:

$$* d * F = \mu_0 j \tag{3.43}$$

Together with the Bianchi identity, Eq. (3.40), this represents all the phenomena of classical electrodynamics.

References

1. P. Szekeres, *A Course in Modern Mathematical Physics: Groups, Hilbert Space and Differential Geometry* (Cambridge University Press, Cambridge, 2004)
2. K. Cahill, *Physical Mathematics*, 2nd edn. (Cambridge University Press, Cambridge, United Kingdom, 2019)

Chapter 4
Differential Calculus on Curved Manifolds

Abstract In this chapter we present an introduction to differential operations on curved manifolds. This is done by starting from affine connections, which provide the central tool in order to define derivatives that take into account the fact that the local Cartesian reference frame is effectively "moving" on the surface of the curved manifold. From this fundamental fact, all rigorous definitions of absolute tensor calculus will follow in a straightforward manner. Reference textbooks on this part are Cahill [1] and Carroll [2].

4.1 Affine Connections

On a curved manifold, ordinary partial derivatives fail to produce geometrically meaningful results, because the local reference frame itself varies from point to point. Affine connections provide the mathematical structure needed to compare vectors defined at nearby points and to define derivatives that respect the geometry of the manifold.

In what follows, the terms "reference frame" or "local Cartesian reference frame" and "local basis" will be used interchangeably.

Given an n-dimensional differentiable manifold $\mathcal{M}$ and a vector field $\mathbf{F}(\mathbf{x})$ on it, fixing a reference frame (RF) $\{\mathbf{e}_i(\mathbf{x})\}_{i=1,\dots,n}$, in computing the derivative of $\mathbf{F}(\mathbf{x})$ one must also take into account how the basis itself varies in space:

$$\frac{\partial \mathbf{F}}{\partial x^j} = \frac{\partial F^i}{\partial x^j}\mathbf{e}_i + F^i \frac{\partial \mathbf{e}_i}{\partial x^j} \tag{4.1}$$

To treat this problem formally, it is necessary to introduce the tangent space, which provides the natural setting in which vectors and their derivatives are defined on a manifold.

A. Zaccone, *Differential Geometry and Group Theory*,
https://doi.org/10.1007/978-3-032-27518-9_4

Definition 4.1 Given a differentiable manifold $\mathcal{M}$ and a point $p \in \mathcal{M}$, the *tangent space* $T_p\mathcal{M}$ is defined as the set of equivalence classes of smooth curves $f : I \subseteq \mathbb{R} \to \mathcal{M}$ such that $f(\lambda_0) = p$, where two curves are equivalent if their coordinate velocities agree at λ_0.

In essence, $T_p\mathcal{M}$ is the space of all tangent vectors to $\mathcal{M}$ based at p.

Definition 4.2 Given a differentiable manifold $\mathcal{M}$, fixing a RF $\{\mathbf{e}_j(\mathbf{x})\}_{j=1,\ldots,n}$ on it, one defines the connection coefficients associated with the coordinate basis (*Christoffel symbol of the second kind*) as:

$$\Gamma^k_{li} \equiv \mathbf{e}^k \cdot \frac{\partial \mathbf{e}_i}{\partial x^l} \tag{4.2}$$

This affine connection relates tangent spaces at two infinitesimally close points. Recalling the derivative of a vector field:

$$\mathbf{e}^k \cdot \frac{\partial \mathbf{F}}{\partial x^l} = \frac{\partial F^i}{\partial x^l} \underbrace{\mathbf{e}^k \cdot \mathbf{e}_i}_{\delta^k_i} + F^i \mathbf{e}^k \cdot \frac{\partial \mathbf{e}_i}{\partial x^l} = \frac{\partial F^k}{\partial x^l} + \Gamma^k_{li} F^i \equiv \nabla_l F^k \tag{4.3}$$

This is defined as the *covariant derivative* of the vector field F^k (with a slight abuse of terminology).

Definition 4.3 A connection is said to be *torsion-free* if

$$\Gamma^k_{li} = \Gamma^k_{il}.$$

If, in addition, it satisfies metric compatibility,

$$\nabla_k g_{ij} = 0,$$

then it is called the *Levi-Civita connection*.

This holds under the hypotheses of Schwarz's lemma, since $\mathbf{e}_i = \frac{\partial \mathbf{p}}{\partial x^i}$:

$$\Gamma^k_{li} = \mathbf{e}^k \cdot \frac{\partial \mathbf{e}_i}{\partial x^l} = \mathbf{e}^k \cdot \frac{\partial^2 \mathbf{p}}{\partial x^l \partial x^i} = \mathbf{e}^k \cdot \frac{\partial^2 \mathbf{p}}{\partial x^i \partial x^l} = \mathbf{e}^k \frac{\partial \mathbf{e}_l}{\partial x^i} = \Gamma^k_{il} \tag{4.4}$$

Under a change of reference frame $x \mapsto y$, the affine connection transforms as:

$$\tilde{\Gamma}^k_{li} = \tilde{\mathbf{e}}^k \cdot \frac{\partial \tilde{\mathbf{e}}_i}{\partial y^l} = \frac{\partial y^k}{\partial x^p} \mathbf{e}^p \frac{\partial x^m}{\partial y^l} \frac{\partial}{\partial x^m} \left(\frac{\partial x^n}{\partial y^i} \mathbf{e}_n \right) = \frac{\partial y^k}{\partial x^p} \frac{\partial x^m}{\partial y^l} \frac{\partial x^n}{\partial y^i} \Gamma^p_{mn} + \frac{\partial y^k}{\partial x^p} \frac{\partial^2 x^p}{\partial y^l \partial y^i} \tag{4.5}$$

The second term shows that, in general, the Levi-Civita affine connection is not a tensor, since it transforms affinely, not homogeneously. This reflects the fact that the affine connection depends explicitly on the choice of reference frame, rather than representing a geometric object intrinsic to the manifold.

In general, the affine connection has n^3 independent components. If the connection is torsion-free, the symmetry $\Gamma^k_{li} = \Gamma^k_{il}$ reduces the number of independent components to

$$n \times \frac{n(n+1)}{2}.$$

In four dimensions this gives $4 \times 10 = 40$ independent components. In flat Minkowski spacetime with Cartesian coordinates, however, all Christoffel symbols vanish identically.

Moreover, the symmetry $\Gamma^k_{li} = \Gamma^k_{il}$ is linked to two important properties:

- preservation of the metric: $\nabla_l g_{mn} = 0$;
- absence of torsion:

$$T(X, Y) \equiv \nabla_X Y - \nabla_Y X - [X, Y] = 0, \quad \forall X, Y \text{ vector fields on } \mathcal{M}.$$

Here the Lie brackets [,] have been defined.

When the above symmetry is not present (e.g. in Einstein–Cartan theory of gravitation), one can cancel the non-homogeneous term in Eq. (4.5) by defining a new tensor, called the torsion tensor:

Definition 4.4 The *torsion tensor* is defined as $T^a_{bc} \equiv \Gamma^a_{bc} - \Gamma^a_{cb}$.

4.2 Parallel Transport

Given an n-dimensional differentiable manifold $\mathcal{M}$ in an embedding space identified with $\mathbb{R}^{n+k}$, with $k \geq 1$, the vectors in the tangent space $T_p\mathcal{M}$ can be viewed as vectors in the embedding space.

Affine connections acquire direct geometric meaning through the notion of parallel transport. Intuitively, one may picture the parallel transport of a vector along a curve on the manifold as moving the application point of the vector on the hypersurface determined by $\mathcal{M}$, while keeping the vector tangent to the manifold and without modifying its components in the embedding space. This leads, if one considers a loop on the manifold, to obtaining a vector oriented differently from the initial one (see Fig. 4.1).

Proposition 4.1 *The condition for the parallel transport of a vector field* $\mathbf{F}(\mathbf{x})$ *is that along the considered curve one has* $\mathbf{e}^k \cdot d\mathbf{F} = 0$.

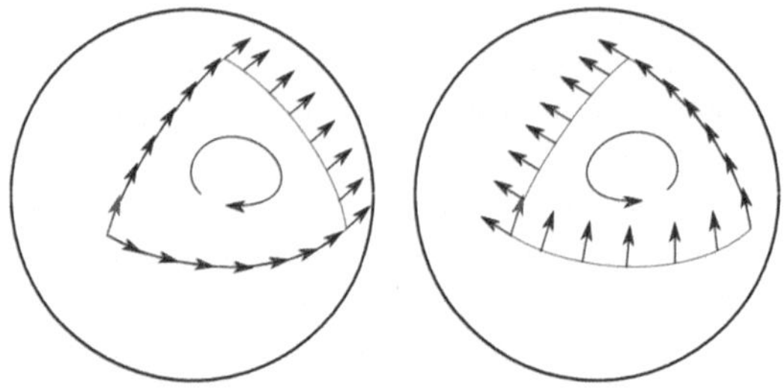

Fig. 4.1 Parallel transport of a vector on closed loops on a sphere. In each case, the orientation of the vector changes after a full loop on the spherical surface. The extent of the change in the vector orientation after the loop is proportional to the area of curved surface enclosed by the loop

Since $d\mathbf{F} = \partial_l \mathbf{F}\, dx^l$, this condition is equivalent to requiring that

$$\nabla_l F^k = 0$$

along the curve.

4.3 Covariant Derivative

The covariant derivative combines ordinary differentiation with the affine connection to produce a derivative operator that transforms tensorially under changes of reference frame.

We introduce the following notation for derivatives: $f_{,i} \equiv \frac{\partial f}{\partial x^i}$, regardless of the nature of f.

Considering a vector field $\mathbf{F}(\mathbf{x})$ on a differentiable manifold $\mathcal{M}$, one can rewrite the covariant derivative as:

$$\nabla_l F^k \equiv F^k_{;l} \equiv \mathbf{e}^k \cdot \mathbf{F}_{,l} = \mathbf{e}^k \cdot \left(F^i_{,l} \mathbf{e}_i + F^i \mathbf{e}_{i,l} \right) = F^k_{,l} + \Gamma^k_{li} F^i \tag{4.6}$$

Proposition 4.2 $\mathbf{e}_k \cdot \mathbf{e}^i_{,l} = -\Gamma^i_{lk}$.

Proof

$$0 = \delta^i_{k,l} = (\mathbf{e}_k \cdot \mathbf{e}^i)_{,l} = \mathbf{e}_{k,l} \cdot \mathbf{e}^i + \mathbf{e}_k \cdot \mathbf{e}^i_{,l} = \Gamma^i_{lk} + \mathbf{e}_k \cdot \mathbf{e}^i_{,l}.$$

□

Thus, the covariant derivative can be expressed both for contravariant and covariant vector fields:

$$V^k_{;l} \equiv \mathbf{e}^k \cdot \mathbf{V}_{,l} = \mathbf{e}^k \cdot \left(V^i_{,l}\mathbf{e}_i + V^i \mathbf{e}_{i,l} \right) \tag{4.7}$$

$$V_{k;l} \equiv \mathbf{e}_k \cdot \mathbf{V}_{,l} = \mathbf{e}_k \cdot \left(V_{i,l}\mathbf{e}^i + V_i \mathbf{e}^i_{,l} \right) \tag{4.8}$$

Expressed in terms of affine connections:

$$\nabla_l V^k = \frac{\partial V^k}{\partial x^l} + \Gamma^k_{li} V^i \tag{4.9}$$

$$\nabla_l V_k = \frac{\partial V_k}{\partial x^l} - \Gamma^i_{lk} V_i \tag{4.10}$$

On torsion-free manifolds ($T^a_{bc} \equiv 0$), for a covariant vector field one has

$$V_{l;i} - V_{i;l} = V_{l,i} - V_{i,l},$$

since the symmetric Christoffel symbols cancel in the antisymmetrization.

Although the affine connection itself is not a tensor, its non-tensorial transformation properties are precisely what ensure that the covariant derivative transforms tensorially.

Proposition 4.3 *The covariant derivative transforms as a tensor.*

We now verify explicitly that the covariant derivative defined above transforms tensorially.

Proof Recalling Eq. (4.5):

$$\begin{aligned}
\tilde{\nabla}_l \tilde{V}^k &= \frac{\partial \tilde{V}^k}{\partial y^l} + \tilde{\Gamma}^k_{li} \tilde{V}^i \\
&= \frac{\partial x^m}{\partial y^l} \frac{\partial}{\partial x^m} \left(\frac{\partial y^k}{\partial x^p} V^p \right) + \tilde{\Gamma}^k_{lj} \tilde{V}^j \\
&= \frac{\partial x^m}{\partial y^l} \frac{\partial y^k}{\partial x^p} \frac{\partial V^p}{\partial x^m} + \frac{\partial x^m}{\partial y^l} \frac{\partial^2 y^k}{\partial x^m \partial x^n} V^n + \frac{\partial y^k}{\partial x^p} \frac{\partial x^m}{\partial y^l} \Gamma^p_{mn} V^n \\
&\quad + \frac{\partial y^k}{\partial x^p} \frac{\partial^2 x^p}{\partial y^l \partial y^j} \frac{\partial y^j}{\partial x^n} V^n \\
&= \frac{\partial x^m}{\partial y^l} \frac{\partial y^k}{\partial x^p} \left(\frac{\partial V^p}{\partial x^m} + \Gamma^p_{mn} V^n \right) + \left(\frac{\partial x^m}{\partial y^l} \frac{\partial^2 y^k}{\partial x^m \partial x^n} + \frac{\partial y^k}{\partial x^p} \frac{\partial^2 x^p}{\partial y^l \partial y^j} \frac{\partial y^j}{\partial x^n} \right) V^n.
\end{aligned}$$

One needs to show that the second term vanishes:

$$\frac{\partial}{\partial y^l}\frac{\partial y^k}{\partial x^n}+\frac{\partial y^k}{\partial x^p}\left[\frac{\partial y^j}{\partial x^n}\frac{\partial}{\partial y^l}\frac{\partial x^p}{\partial y^j}\right]=$$

$$=\frac{\partial}{\partial y^l}\frac{\partial y^k}{\partial x^n}+\frac{\partial y^k}{\partial x^p}\left[\frac{\partial}{\partial y^l}\left(\frac{\partial y^j}{\partial x^n}\frac{\partial x^p}{\partial y^j}\right)-\frac{\partial x^p}{\partial y^j}\frac{\partial}{\partial y^l}\frac{\partial y^j}{\partial x^n}\right]$$

$$=\frac{\partial}{\partial y^l}\frac{\partial y^k}{\partial x^n}+\frac{\partial y^k}{\partial x^p}\frac{\partial}{\partial y^l}\delta^p_n-\delta^k_j\frac{\partial}{\partial y^l}\frac{\partial y^j}{\partial x^n}$$

$$=\frac{\partial}{\partial y^l}\frac{\partial y^k}{\partial x^n}-\frac{\partial}{\partial y^l}\frac{\partial y^k}{\partial x^n}=0.$$

The case of the covariant derivative of a covariant vector is analogous. □

Proposition 4.4 *The covariant derivative of a rank-2 tensor is:*

$$\nabla_k A_{ij}=\frac{\partial A_{ij}}{\partial x^k}-\Gamma^m_{ik}A_{mj}-\Gamma^m_{jk}A_{im} \tag{4.11}$$

$$\nabla_k A^{ij}=\frac{\partial A^{ij}}{\partial x^k}+\Gamma^i_{mk}A^{mj}+\Gamma^j_{mk}A^{im} \tag{4.12}$$

Proof Let $\mathbf{A}=A_{ij}\,\mathbf{e}^i\otimes\mathbf{e}^j$, then:

$$\frac{\partial\mathbf{A}}{\partial x^k}=\frac{\partial A_{ij}}{\partial x^k}\,\mathbf{e}^i\otimes\mathbf{e}^j+A_{ij}\frac{\partial\mathbf{e}^i}{\partial x^k}\otimes\mathbf{e}^j+A_{ij}\mathbf{e}^i\otimes\frac{\partial\mathbf{e}^j}{\partial x^k} \tag{4.13}$$

$$=\frac{\partial A_{ij}}{\partial x^k}\,\mathbf{e}^i\otimes\mathbf{e}^j-A_{ij}\Gamma^i_{mk}\mathbf{e}^m\otimes\mathbf{e}^j-A_{ij}\Gamma^j_{mk}\mathbf{e}^i\otimes\mathbf{e}^m \tag{4.14}$$

$$=\frac{\partial A_{ij}}{\partial x^k}\,\mathbf{e}^i\otimes\mathbf{e}^j-A_{mj}\Gamma^m_{ik}\mathbf{e}^i\otimes\mathbf{e}^j-A_{im}\Gamma^m_{jk}\mathbf{e}^i\otimes\mathbf{e}^j \tag{4.15}$$

$$=\left(\frac{\partial A_{ij}}{\partial x^k}-\Gamma^m_{ik}A_{mj}-\Gamma^m_{jk}A_{im}\right)\mathbf{e}^i\otimes\mathbf{e}^j \tag{4.16}$$

$$\equiv(\nabla_k A_{ij})\,\mathbf{e}^i\otimes\mathbf{e}^j. \tag{4.17}$$

□

Proposition 4.5 *Given A_{ij}, an antisymmetric rank-2 tensor on a torsion-free manifold, one has:*

$$A_{ij;k}+A_{ki;j}+A_{jk;i}=A_{ij,k}+A_{ki,j}+A_{jk,i}. \tag{4.18}$$

Proof From Proposition 4.4:

$$A_{ij;k} + A_{ki;j} + A_{jk;i} = A_{ij,k} - \Gamma^m_{ik} A_{mj} - \Gamma^m_{jk} A_{im} + A_{ki,j} - \Gamma^m_{kj} A_{mi} - \Gamma^m_{ij} A_{km} + A_{jk,i} - \Gamma^m_{ji} A_{mk} - \Gamma^m_{ki} A_{jm}.$$

By the symmetry of Γ^m_{ik} and the antisymmetry of A_{ij}:

$$\Gamma^m_{ik} A_{mj} = -\Gamma^m_{ki} A_{jm}, \qquad \Gamma^m_{jk} A_{im} = -\Gamma^m_{kj} A_{mi}, \qquad \Gamma^m_{ij} A_{km} = -\Gamma^m_{ji} A_{mk}.$$

Hence the claim follows. □

Thus, the Bianchi identity holds in any reference frame on a torsion-free manifold.

4.4 Torsion-Free Manifolds

4.4.1 Affine Connections

Definition 4.5 The *Christoffel symbols of the first kind* are defined as

$$\Gamma_{ijk} \equiv g_{im} \Gamma^m_{jk}. \tag{4.19}$$

Proposition 4.6 *On a torsion-free manifold,* $\Gamma_{ijk} = \Gamma_{ikj}$.

Proof This is immediate, recalling that on a torsion-free manifold $\Gamma^m_{jk} = \Gamma^m_{kj}$. □

Proposition 4.7 $\Gamma^m_{ij} = g^{km} \Gamma_{kij}$.

Proof $g^{km} \Gamma_{kij} = g^{km} g_{kn} \Gamma^n_{ij} = \delta^m_n \Gamma^n_{ij} = \Gamma^m_{ij}$. □

It is possible to express affine connections in terms of the metric tensor.

Proposition 4.8

$$\Gamma_{ijk} = \frac{1}{2} \left(g_{ij,k} + g_{ki,j} - g_{jk,i} \right). \tag{4.20}$$

Proof First note that

$$\Gamma_{ijk} = g_{im} \Gamma^m_{jk} = g_{im} \mathbf{e}^m \cdot \mathbf{e}_{k,j} = \mathbf{e}_i \cdot \mathbf{e}_{k,j}.$$

Consequently:

$$g_{ij,k} = \mathbf{e}_{i,k} \cdot \mathbf{e}_j + \mathbf{e}_i \cdot \mathbf{e}_{j,k} = \Gamma_{jki} + \Gamma_{ikj},$$

and similarly

$$g_{ki,j} = \Gamma_{ijk} + \Gamma_{kji}, \qquad g_{jk,i} = \Gamma_{kij} + \Gamma_{jik}.$$

Thus:

$$\begin{aligned} g_{ij,k} + g_{ki,j} - g_{jk,i} &= \Gamma_{jki} + \Gamma_{ikj} + \Gamma_{ijk} + \Gamma_{kji} - \Gamma_{kij} - \Gamma_{jik} \\ &= \Gamma_{jik} - \Gamma_{ijk} + \Gamma_{ijk} + \Gamma_{kij} - \Gamma_{kij} - \Gamma_{jik} \\ &= 2\Gamma_{ijk}. \end{aligned}$$

□

Proposition 4.9

$$\Gamma^m_{ij} = \frac{1}{2} g^{km} \left(g_{mi,j} + g_{jm,i} - g_{ij,m} \right). \tag{4.21}$$

Proof From Propositions 4.7 and 4.8. □

4.4.2 Covariant Derivative of the Metric Tensor

From Proposition 4.4 one has:

$$\nabla_k g_{ij} = \frac{\partial g_{ij}}{\partial x^k} - \Gamma^m_{ik} g_{mj} - \Gamma^m_{jk} g_{im}. \tag{4.22}$$

Thus:

$$\begin{aligned} g_{ij;k} &= g_{ij,k} - \Gamma_{jik} - \Gamma_{ijk} \\ &= g_{ij,k} - \tfrac{1}{2}\left(g_{ji,k} + g_{kj,i} - g_{ik,j}\right) - \tfrac{1}{2}\left(g_{ij,k} + g_{ki,j} - g_{jk,i}\right) \\ &= g_{ij,k} - \tfrac{1}{2} g_{ij,k} + \tfrac{1}{2} g_{jk,i} + \tfrac{1}{2} g_{ik,j} - \tfrac{1}{2} g_{ij,k} - \tfrac{1}{2} g_{ik,j} + \tfrac{1}{2} g_{jk,i} \\ &= 0. \end{aligned}$$

We thus see one of the fundamental properties of torsion-free manifolds, namely the preservation of the metric:

$$\nabla_k g_{ij} = 0. \tag{4.23}$$

This condition expresses the compatibility of the Levi-Civita connection with the metric and guarantees that lengths and angles are preserved under parallel transport.

4.5 Differential Operators on Curved Manifolds

Using the covariant derivative, it is now possible to generalize the familiar differential operators of vector calculus—gradient, divergence, and Laplacian—to curved manifolds.

Given a generic n-dimensional differentiable manifold $\mathcal{M}$, fixing an orthogonal RF $\{\mathbf{e}_i\}_{i=1,\ldots,n}$, it is possible to define on $\mathcal{M}$ the usual differential operators of Euclidean space.

4.5.1 *Gradient*

Definition 4.6 Given a scalar field $f : \mathcal{M} \to \mathbb{R}$, its differential is defined as:

$$df(\mathbf{x}) = f(\mathbf{x}+d\mathbf{x}) - f(\mathbf{x}) = \frac{\partial f}{\partial x^i} dx^i \equiv \nabla f(\mathbf{x}) \cdot d\mathbf{x}. \tag{4.24}$$

Proposition 4.10

$$\nabla f = \sum_{i=1}^{n} \frac{1}{h_i} \frac{\partial f}{\partial x^i} \hat{\mathbf{e}}_i .$$

Proof Since $\frac{\hat{\mathbf{e}}_i}{h_i} = \mathbf{e}^i$ (correctly, $\frac{\partial f}{\partial x^i}$ is covariant):

$$\nabla f \cdot d\mathbf{x} = \frac{\partial f}{\partial x^i} \mathbf{e}^i \cdot dx^j \mathbf{e}_j = \frac{\partial f}{\partial x^i} dx^j \delta^j_i = \frac{\partial f}{\partial x^i} dx^i = df.$$

□

4.5.2 *Divergence*

Definition 4.7 Given a contravariant vector field $\mathbf{V} = V^i \mathbf{e}_i$, its divergence is defined as:

$$\nabla \cdot \mathbf{V} \equiv V^i_{;i} = V^i_{,i} + \Gamma^i_{ik} V^k. \tag{4.25}$$

The divergence is nothing but a contraction of the covariant derivative.

Proposition 4.11

$$\Gamma^i_{ik} = \tfrac{1}{2} g^{im} g_{im,k}.$$

Proof

$$\Gamma^i_{ik} = \tfrac{1}{2} g^{im} \big(g_{im,k} + g_{ki,m} - g_{mk,i} \big).$$

But $g^{im} g_{mk,i} = g^{mi} g_{ik,m}$ since m, i are dummy indices, hence the claim. □

It is possible to give an alternative formulation. Let c denote the cofactor matrix of the metric tensor g (viewed as a matrix):

$$\det g = \sum_{i,j} g_{ij} c_{ij}, \qquad \Longrightarrow \qquad c_{ij} = \frac{\partial \det g}{\partial g_{ij}}. \tag{4.26}$$

Recalling that the inverse metric is

$$g^{ij} = \frac{1}{\det g} \frac{\partial \det g}{\partial g_{ij}}, \tag{4.27}$$

Proposition 4.12

$$\Gamma^i_{ik} = \frac{(\sqrt{g})_{,k}}{\sqrt{g}}.$$

Proof First,

$$(\det g)_{,k} = g_{ij,k} \frac{\partial \det g}{\partial g_{ij}} = g_{ij,k} g^{ij} \det g.$$

Then, from Proposition 4.11:

$$\Gamma^i_{ik} = \tfrac{1}{2} g^{im} g_{im,k} = \tfrac{1}{2} \frac{1}{\det g} (\det g)_{,k} = \tfrac{1}{2|\det g|} |\det g|_{,k} = \frac{(\sqrt{g})_{,k}}{\sqrt{g}}.$$

□

Proposition 4.13 *Given a contravariant vector field* $\mathbf{V} = V^i \mathbf{e}_i$*, its divergence is:*

$$\nabla \cdot \mathbf{V} = \frac{(\sqrt{g}\, V^k)_{,k}}{\sqrt{g}}. \tag{4.28}$$

Proof From Proposition 4.12:

$$\nabla \cdot \mathbf{V} = V^i_{,i} + \Gamma^i_{ik} V^k = V^i_{,i} + \frac{(\sqrt{g})_{,k}}{\sqrt{g}} V^k = \frac{(\sqrt{g}\, V^k)_{,k}}{\sqrt{g}}.$$

□

4.5.2.1 Example: Divergence in Orthogonal Coordinates

In $2d$ with orthogonal coordinates, we recall that

$$g_{ij} = e_i \cdot e_j = h_i^2 \delta_{ij}, \tag{4.29}$$

$$V_i \equiv h_i \hat{V}^i = h_i^{-1} V^i \quad \text{(no sum here)} \tag{4.30}$$

and

$$\mathbf{V} = \hat{e}_i V^i, \tag{4.31}$$

where $\hat{e}_i$ are orthonormal.

These conditions imply that:

$$\sqrt{g} = h_1 h_2 \quad \left(\sqrt{g} = h_1 h_2 \cdots h_n \text{ in } n \text{ dimensions}\right) \tag{4.32}$$

with

$$V^k = \frac{\bar{V}_k}{h_k}$$

and therefore, the divergence of a vector $\mathbf{V}$ is given by:

$$\nabla \cdot \mathbf{V} = \frac{1}{h_1 h_2} \sum_{k=1}^{2} \left(\frac{h_1 h_2}{h_k} \bar{V}^k \right)_{,k} \tag{4.33}$$

Compare this expression with:

$$\nabla \cdot \mathbf{V} = \frac{(\sqrt{g}\, V^k)_{,k}}{\sqrt{g}}$$

where

$$V^k = \frac{\bar{V}_k}{h_k}.$$

In polar coordinates in $2d$: $h_r = 1$, $h_\theta = r$, hence

$$\nabla \cdot \mathbf{V} = \frac{1}{r}\Big[(r V_r)_{,r} + (V_\theta)_{,\theta}\Big] = \frac{1}{r}(r V_r)_{,r} + \frac{1}{r} V_{\theta,\theta}.$$

In $3d$ we have

$$\sqrt{g} = h_1 h_2 h_3, \quad V^k = \frac{\bar{V}_k}{h_k},$$

and

$$\nabla \cdot \mathbf{V} = \frac{1}{h_1 h_2 h_3} \sum_{k=1}^{3} \left(\frac{h_1 h_2 h_3}{h_k} \bar{V}^k \right)_{,k}.$$

In cylindrical coordinates: $h_\rho = 1, h_\phi = \rho, h_z = 1$

$$\nabla \cdot \mathbf{V} = \frac{1}{\rho} \Big[(\rho V_\rho)_{,\rho} + (V_\phi)_{,\phi} + \rho (V_z)_{,z} \Big]$$

$$= \frac{1}{\rho} \Big[(\rho V_\rho)_{,\rho} \Big] + \frac{1}{\rho} V_{\phi,\phi} + V_{z,z}.$$

In spherical coordinates: $h_r = 1, h_\theta = r, h_\phi = r \sin\theta$

$$g = |\det g_{ij}| = r^4 \sin^2\theta,$$

and therefore the metric tensor is:

$$\left(g^{ij}\right) = \begin{pmatrix} 1 & 0 & 0 \\ 0 & r^{-2} & 0 \\ 0 & 0 & (r^2 \sin^2\theta)^{-1} \end{pmatrix},$$

from which:

$$\nabla \cdot \mathbf{V} = \frac{1}{r^2 \sin\theta} \Big[\left(r^2 \sin\theta \, \bar{V}_r\right)_{,r} + \left(r \sin\theta \, \bar{V}_\theta\right)_{,\theta} + \left(r \, \bar{V}_\phi\right)_{,\phi} \Big]$$

$$= \frac{1}{r^2 \sin\theta} \Big[(\sin\theta \, (r^2 \bar{V}_r))_{,r} + r \, (\sin\theta \, \bar{V}_\theta)_{,\theta} + r \, \bar{V}_{\phi,\phi} \Big].$$

4.5.3 Generalized Kronecker Delta

Definition 4.8 The *generalized Kronecker delta* is defined as:

$$\delta^{\mu_1 \ldots \mu_k}_{\nu_1 \ldots \nu_k} \equiv \begin{cases} +1 & (\nu_1, \ldots, \nu_k) = \sigma_p(\mu_1, \ldots, \mu_k) \text{ (even permutation)} \\ -1 & (\nu_1, \ldots, \nu_k) = \sigma_d(\mu_1, \ldots, \mu_k) \text{ (odd permutation)} \\ 0 & \text{otherwise} \end{cases} \tag{4.34}$$

Lemma 4.1

$$\delta^{\mu_1 \ldots \mu_k}_{\nu_1 \ldots \nu_k} = \det \begin{bmatrix} \delta^{\mu_1}_{\nu_1} & \ldots & \delta^{\mu_1}_{\nu_k} \\ \vdots & \ddots & \vdots \\ \delta^{\mu_k}_{\nu_1} & \ldots & \delta^{\mu_k}_{\nu_k} \end{bmatrix}. \tag{4.35}$$

Lemma 4.2

$$\epsilon_{i_1 \ldots i_k i_{k+1} \ldots i_n} \, \epsilon^{i_1 \ldots i_k j_{k+1} \ldots j_n} = k! \, \delta^{j_{k+1} \ldots j_n}_{i_{k+1} \ldots i_n}.$$

The language of differential forms provides an alternative and highly compact formulation of differential operators on curved manifolds.

4.5.4 Divergence via Differential Forms

Proposition 4.14 *Given a vector field* $\mathbf{V} = V^i \mathbf{e}_i$, *let* $V = V_i dx^i$ *be the differential form associated to it. Then:*

$$\nabla \cdot \mathbf{V} = \sigma * d * V. \tag{4.36}$$

Proof For simplicity consider the 4-dimensional case (using Lemma 4.2):

$$\begin{aligned}
* d * V &= * d \left(V_i \frac{1}{3!} g^{ik} \varepsilon_{klmn} dx^l \wedge dx^m \wedge dx^n \right) \\
&= * \left[\frac{1}{3!} \left(\sqrt{g} V^k \right)_{,p} \epsilon_{klmn} dx^p \wedge dx^l \wedge dx^m \wedge dx^n \right] \\
&= \frac{1}{3!} \left(\sqrt{g} V^k \right)_{,p} \epsilon_{klmn} g^{p\alpha} g^{l\beta} g^{m\gamma} g^{n\delta} \sqrt{g} \, \epsilon_{\alpha\beta\gamma\delta} \\
&= \frac{1}{3!} \left(\sqrt{g} V^k \right)_{,p} \sqrt{g} \, \epsilon_{klmn} (\det g)^{-1} \epsilon^{plmn} \\
&= \frac{1}{3!} \frac{\sigma}{\sqrt{g}} \left(\sqrt{g} V^k \right)_{,p} (3! \delta^p_k) \\
&= \frac{\sigma}{\sqrt{g}} \left(\sqrt{g} V^k \right)_{,k}.
\end{aligned}$$

□

4.5.5 Example

In a n-dimensional space with $g_{ij} = h_i^2 \delta_{ij}$ one has $\bar{V}_i \equiv h_i^{-1} V_i = h_i V^i$, hence:

$$\nabla \cdot \mathbf{V} = \frac{1}{h_1 \cdots h_n} \sum_{k=1}^{n} \frac{\partial}{\partial x^k} \left(\frac{h_1 \cdots h_n}{h_k} V_k \right).$$

4.5.6 Laplacian

Definition 4.9 Given a scalar field $f : \mathcal{M} \to \mathbb{R}$, its Laplacian is defined as:

$$\Box f \equiv \nabla \cdot \nabla f = \left(g^{ik} f_{,k} \right)_{;i} . \tag{4.37}$$

Proposition 4.15

$$\Box f = \frac{(\sqrt{g} f^{,k})_{,k}}{\sqrt{g}}. \tag{4.38}$$

Proof Directly from Proposition 4.13, using $g^{ik} f_{,k} = f^{,i}$. □

Proposition 4.16 *Given a scalar field $f : \mathcal{M} \to \mathbb{R}$, one has:*

$$\Box f = \sigma * d * df. \tag{4.39}$$

Proof

$$\begin{aligned}
* d * df &= * d \left(f_{,i} * dx^i \right) \\
&= * d \left(\frac{1}{3!} \sqrt{g} f_{,i} g^{ik} \epsilon_{klmn} dx^l \wedge dx^m \wedge dx^n \right) \\
&= * \left(\frac{1}{3!} (\sqrt{g} f^{,k})_{,p} \epsilon_{klmn} dx^p \wedge dx^l \wedge dx^m \wedge dx^n \right) \\
&= \frac{1}{3!} (\sqrt{g} f^{,k})_{,p} \epsilon_{klmn} g^{p\alpha} g^{l\beta} g^{m\gamma} g^{n\delta} \sqrt{g}\, \epsilon_{\alpha\beta\gamma\delta} \\
&= \frac{1}{3!} (\sqrt{g} f^{,k})_{,p} \epsilon_{klmn} \sqrt{g} (\det g)^{-1} \epsilon^{plmn} \\
&= \frac{1}{3!} \frac{\sigma}{\sqrt{g}} (\sqrt{g} f^{,k})_{,p} (3! \delta_k^p) \\
&= \frac{\sigma}{\sqrt{g}} (\sqrt{g} f^{,k})_{,k}.
\end{aligned}$$

□

where we used the identity

$$g^{p\alpha} g^{l\beta} g^{m\gamma} g^{n\delta} \epsilon_{\alpha\beta\gamma\delta} = (\det g)^{-1} \epsilon^{plmn},$$

which follows from standard properties of determinants and the Levi-Civita tensor, and we also used the identity from linear algebra: $|A|\, \epsilon_{mn} = A_{li} A_{mj} A_{nk}\, \epsilon_{ijk}$ where A is a generic matrix.

4.5.6.1 Example

In a n-dimensional space with $g_{ij} = h_i^2 \delta_{ij}$ one has $f^{,k} = g^{ik} f_{,i} = h_k^{-2} f_{,k}$, hence:

$$\nabla^2 f = \frac{1}{h_1 \cdots h_n} \sum_{k=1}^{n} \frac{\partial}{\partial x^k} \left(\frac{h_1 \cdots h_n}{h_k^2} \frac{\partial f}{\partial x^k} \right).$$

4.6 Covariant Electrodynamics Revisited

We now revisit Maxwell's equations, previously introduced in Chap. 3 using differential forms, with the goal of showing explicitly how they can be formulated for curved manifolds.

At first sight, it may appear surprising that curvature does not enter explicitly into Maxwell's equations. This is not an omission, but a fundamental property of classical electrodynamics: in its minimal formulation, the electromagnetic field couples to space-time geometry only through the metric, not through curvature tensors. As a result, the effects of curvature enter implicitly via the Hodge operator and the volume element, while the field strength itself remains defined by exterior differentiation.

The formalism developed in this chapter is therefore essential not because Maxwell's equations require curvature, but because it provides the correct geometric framework in which they remain valid *also* on arbitrary curved manifolds.

As shown in Chap. 3, Maxwell's equations can be written compactly in terms of differential forms as

$$dF = 0, \qquad *d * F = \mu_0 j.$$

The purpose of the present section is to show that they remain valid on curved manifolds, with all metric dependence entering through the Hodge operator and the volume element.

In Minkowski spacetime $g_{ij} = \text{diag}(-1, 1, 1, 1)$, this reduces to $*d * F = \partial_p (F^{kp}) dx_k$. In a generic curved space-time, the same equation holds with the full metric $g_{ij}(x)$ entering through the Hodge dual. Recalling from Eq. (2.28) that

$\partial_p(F^{kp}) = \mu_0 j^k$, and defining the electric current 1-form $j \equiv j^k dx_k$, one obtains the generalization on a curved manifold of the second covariant Maxwell equation:

$$* d * F = \mu_0 j. \tag{4.40}$$

Electromagnetism is, therefore, insensitive to curvature tensors at the level of field equations; it only "feels" geometry through the metric. By contrast, gravitational dynamics explicitly involves curvature, as will be discussed in the next chapter.

References

1. K. Cahill, *Physical Mathematics*, 2nd edn. (Cambridge University Press, Cambridge, United Kingdom, 2019)
2. S.M. Carroll, *Spacetime and Geometry: An Introduction to General Relativity* (Addison-Wesley, San Francisco, 2004)

Chapter 5
General Relativity

Abstract In the previous chapter we have seen how the concept of moving reference frames leads directly to the concept of covariant derivative, by way of defining the affine connections. Then it became straightforward to generalize all differential operations to curved manifolds. In this chapter, these tools will be used to study what happens when an action is varied on a curved manifold. After recalling the basic tenets of the variational principles the tools developed in the previous chapter will be applied to deriving the theory of general relativity, with several physical examples. The central idea is that gravity can be described as a geometric effect of spacetime curvature, emerging naturally from the variational principle once the metric is treated as a dynamical field.

5.1 Stationary Action Principle

In the previous chapter, we developed the differential-geometric framework needed to describe physics on curved manifolds, including covariant derivatives, affine connections, and curvature. We now combine these tools with the principle of stationary action, which provides the dynamical laws governing physical systems.

5.1.1 Classical Case

Definition 5.1 Given a system described by a Lagrangian L, the action is defined as:

$$S[\mathbf{x}(t)] \equiv \int_{t_1}^{t_2} L(\mathbf{x}, \dot{\mathbf{x}})dt. \tag{5.1}$$

A. Zaccone, *Differential Geometry and Group Theory*,
https://doi.org/10.1007/978-3-032-27518-9_5

The principle of least action states that the trajectory followed by the system is an extremal of the action; that is, considering a variation of the trajectory $\delta\mathbf{x} : \delta\mathbf{x}(t_1) = \delta\mathbf{x}(t_2) = \mathbf{0}$:

$$\delta S = 0. \tag{5.2}$$

5.1.1.1 Classical Free Particle

Classically, a free particle is described by $L = \frac{1}{2}m\dot{\mathbf{x}}^2$, hence:

$$\begin{aligned} 0 = \delta S &= \int_{t_1}^{t_2} \delta L \, dt = \int_{t_1}^{t_2} m\dot{\mathbf{x}} \cdot \delta\dot{\mathbf{x}} \, dt = \int_{t_1}^{t_2} \left[m\frac{d}{dt}(\dot{\mathbf{x}} \cdot \delta\mathbf{x}) - m\ddot{\mathbf{x}} \cdot \delta\mathbf{x} \right] dt \\ &= m\left[\ddot{\mathbf{x}} \cdot \delta\mathbf{x}\right]_{t_1}^{t_2} - m\int_{t_1}^{t_2} \ddot{\mathbf{x}} \cdot \delta\mathbf{x} \, dt = -m\int_{t_1}^{t_2} \ddot{\mathbf{x}} \cdot \delta\mathbf{x} \, dt. \end{aligned}$$

Therefore, given the arbitrariness of $\delta\mathbf{x}$, one obtains the equation of motion of the classical free particle:

$$\ddot{\mathbf{x}} = \mathbf{0}. \tag{5.3}$$

5.1.1.2 Motion in a Potential

In the presence of a potential, the Lagrangian becomes $L = \frac{1}{2}m\dot{\mathbf{x}}^2 - V(\mathbf{x})$, so, since $\delta V(\mathbf{x}) = \nabla V(\mathbf{x}) \cdot \delta\mathbf{x}$, the equation of motion is:

$$0 = \delta S = \int_{t_1}^{t_2} (-m\ddot{\mathbf{x}} - \nabla V(\mathbf{x})) \cdot \delta\mathbf{x} \, dt \quad \Longrightarrow \quad m\ddot{\mathbf{x}} = -\nabla V(\mathbf{x}). \tag{5.4}$$

5.1.2 Relativistic Free Particle

While the classical action principle already hints at a geometric interpretation of motion, special relativity makes this structure explicit by introducing spacetime intervals and invariant actions. This prepares the ground for a natural generalization to curved spacetime.

Proposition 5.1 *A relativistic free particle is described by* $L = -\frac{mc^2}{\gamma}$.

Proof Since

$$L = -mc^2\sqrt{1 - \frac{\mathbf{v}^2}{c^2}},$$

we compute

$$\frac{\partial}{\partial v_i}\sqrt{1-\frac{\mathbf{v}^2}{c^2}} = \frac{-v_i/c^2}{\sqrt{1-\frac{\mathbf{v}^2}{c^2}}} = -\frac{\gamma v_i}{c^2}.$$

Hence

$$p_i = \frac{\partial L}{\partial v_i} = -mc^2\left(-\frac{\gamma v_i}{c^2}\right) = \gamma m v_i,$$

which is the correct relativistic momentum. □

The action describing a relativistic free particle is:

$$S = -mc^2 \int_{t_1}^{t_2} \frac{dt}{\gamma} = -mc \int_{\tau_1}^{\tau_2} d\tau. \tag{5.5}$$

Setting $c = 1$, one can compute the equations of motion:

$$\begin{aligned}
0 = \delta S &= -\delta \int_{t_1}^{t_2} m\sqrt{1-\dot{\mathbf{x}}^2}\, dt = m \int_{t_1}^{t_2} \frac{\dot{\mathbf{x}}\cdot\delta\dot{\mathbf{x}}}{\sqrt{1-\dot{\mathbf{x}}^2}} dt = m \int_{t_1}^{t_2} \frac{d\mathbf{x}}{dt}\cdot\frac{d\delta\mathbf{x}}{dt}\frac{dt}{d\tau} dt \\
&= m \int_{\tau_1}^{\tau_2} \frac{d\mathbf{x}}{dt}\cdot\frac{d\delta\mathbf{x}}{dt}\frac{dt}{d\tau}\frac{dt}{d\tau} d\tau = m \int_{\tau_1}^{\tau_2} \frac{d\mathbf{x}}{d\tau}\cdot\frac{d\delta\mathbf{x}}{d\tau} d\tau \\
&= m \int_{\tau_1}^{\tau_2} \left[\frac{d}{d\tau}(\dot{\mathbf{x}}\cdot\delta\mathbf{x}) - \frac{d^2\mathbf{x}}{d\tau^2}\cdot\delta\mathbf{x}\right] d\tau = -m \int_{\tau_1}^{\tau_2} \frac{d^2\mathbf{x}}{d\tau^2}\cdot\delta\mathbf{x}\, d\tau.
\end{aligned}$$

From the arbitrariness of $\delta\mathbf{x}$ one obtains the equation of motion:

$$\frac{d^2\mathbf{x}}{d\tau^2} = \mathbf{0}. \tag{5.6}$$

5.1.2.1 Electromagnetic Field

To describe the motion of a particle in an electromagnetic field, one must add an interaction term to the Lagrangian:

$$S = -m \int_{\tau_1}^{\tau_2} d\tau + q \int_{\mathbf{x}_1}^{\mathbf{x}_2} A_i(\mathbf{x}) dx^i = \int_{\tau_1}^{\tau_2} \left(-m + qA_i(\mathbf{x})\frac{dx^i}{d\tau}\right) d\tau. \tag{5.7}$$

To obtain the equations of motion, therefore:

$$
\begin{aligned}
0 = \delta S &= -m \int_{\tau_1}^{\tau_2} \delta\sqrt{-\eta_{ij}dx^i dx^j} + q \int_{\mathbf{x}_1}^{\mathbf{x}_2} \delta\left(A_i dx^i\right) \\
&= -m \int_{\tau_1}^{\tau_2} \frac{1}{2}\frac{\delta\left(-\eta_{ij}dx^i dx^j\right)}{\sqrt{-\eta_{ij}dx^i dx^j}} + q \int_{\mathbf{x}_1}^{\mathbf{x}_2} \left(\delta A_i dx^i + A_i \delta dx^i\right) \\
&= m \int_{\tau_1}^{\tau_2} \frac{1}{2}\frac{\eta_{ij}\left(\delta dx^i dx^j + dx^i \delta dx^j\right)}{\sqrt{-\eta_{ij}dx^i dx^j}} + q \int_{\tau_1}^{\tau_2} \left(\delta A_i \frac{dx^i}{d\tau} + A_i \frac{d\delta x^i}{d\tau}\right) d\tau \\
&= \int_{\tau_1}^{\tau_2} \left(m \frac{\eta_{ij}dx^i}{\sqrt{-\eta_{ij}dx^i dx^j}} \frac{d\delta x^j}{d\tau} + q \frac{\partial A_i}{\partial x^k}\delta x^k \frac{dx^i}{d\tau} + q A_i \frac{d\delta x^i}{d\tau} \right) d\tau \\
&= \int_{\tau_1}^{\tau_2} \left(m\eta_{ij}\frac{dx^i}{d\tau}\frac{d\delta x^j}{d\tau} + q A_{i,k}\delta x^k \frac{dx^i}{d\tau} + q A_i \frac{d\delta x^i}{d\tau} \right) d\tau \\
&= \int_{\tau_1}^{\tau_2} \left(-m \frac{du_k}{d\tau}\delta x^k + q A_{i,k} u^i \delta x^k - q \frac{dA_k}{d\tau}\delta x^k \right) d\tau \\
&= \int_{\tau_1}^{\tau_2} \left(-m \frac{du_k}{d\tau} + q\left(A_{i,k} - A_{k,i}\right) u^i \right) \delta x^k d\tau
\end{aligned} \tag{5.8}
$$

where the four-velocity u_k has been used. Recalling the definition of the Faraday tensor $F_{ij} \equiv A_{j,i} - A_{i,j}$, one obtains the covariant form of the Lorentz force:

$$
\frac{dp_k}{d\tau} = q F_{ki} u^i . \tag{5.9}
$$

This confirms the choice of Lagrangian.

5.1.3 Gravitational Field

The transition from special to general relativity is achieved by allowing the spacetime metric to become position-dependent. In this way, gravitational effects are encoded in the geometry of spacetime rather than in an external force field.

Consider a particle on a curved manifold:

$$
0 = \delta S = -m \int_{\tau_1}^{\tau_2} \delta d\tau = -m \int_{\tau_1}^{\tau_2} \delta\sqrt{-g_{ij}dx^i dx^j} \tag{5.10}
$$

$$= m\int_{\tau_1}^{\tau_2} \frac{1}{2}\frac{\delta g_{ij}dx^i dx^j + 2g_{ij}dx^i \delta dx^j}{\sqrt{-g_{ij}dx^i dx^j}} = m\int_{\tau_1}^{\tau_2}\left(\frac{1}{2}g_{ij,k}u^i u^j \delta x^k d\tau + g_{ij}u^i \delta dx^j\right) \tag{5.11}$$

$$= m\int_{\tau_1}^{\tau_2}\left(\frac{1}{2}g_{ij,k}u^i u^j \delta x^k + g_{ij}u^i \frac{d\delta x^j}{d\tau}\right)d\tau \tag{5.12}$$

$$= m\int_{\tau_1}^{\tau_2}\left(\frac{1}{2}g_{ij,k}u^i u^j \delta x^k - \frac{d}{d\tau}\left(g_{ij}u^i\right)\delta x^j\right)d\tau \tag{5.13}$$

$$= m\int_{\tau_1}^{\tau_2}\left(\frac{1}{2}g_{ij,k}u^i u^j - \frac{d}{d\tau}\left(g_{ik}u^i\right)\right)\delta x^k d\tau. \tag{5.14}$$

Since δx^k is arbitrary, one obtains:

$$\frac{1}{2}g_{ij,k}u^i u^j - g_{ik,j}u^i u^j - g_{ik}\frac{du^i}{d\tau} = 0 \tag{5.15}$$

Multiplying through by g^{rk}:

$$\frac{du^r}{d\tau} + g^{rk}\left(g_{ik,j} - \frac{1}{2}g_{ij,k}\right)u^i u^j = 0. \tag{5.16}$$

Since only the part symmetric in i, j survives when multiplied by $u^i u^j$:

$$\frac{du^r}{d\tau} + \frac{1}{2}g^{rk}\left(g_{ik,j} + g_{jk,i} - g_{ij,k}\right)u^i u^j = 0. \tag{5.17}$$

One recognizes the definition of the Levi-Civita connection (Eq. (4.9)):

$$\frac{d^2x^r}{d\tau^2} + \Gamma^r_{ij}\frac{dx^i}{d\tau}\frac{dx^j}{d\tau} = 0. \tag{5.18}$$

This is known as the *geodesic equation* and describes the motion of a free particle on a curved manifold: one sees that, in addition to the inertial term, there is a force term due to the very geometry of space; moreover, one can observe that the trajectory is independent of the mass of the body. Although the Levi-Civita connection is not a tensor, one can show that the force term cancels the inhomogeneous terms, making the geodesic equation a tensorial equation. Thus, the variational principle identifies free-fall motion with geodesics of the spacetime metric, establishing the geometric nature of gravity.

In general, solutions of the geodesic equation are called geodesics.

Definition 5.2 Given a differentiable manifold $\mathcal{M}$ with Levi-Civita connection ∇, a *geodesic* is a curve γ whose tangent vector $\dot{\gamma}$ is parallel transported along itself:

$$\nabla_{\dot{\gamma}}\dot{\gamma} = 0.$$

Equivalently, a geodesic is a stationary curve of the length functional.

Proposition 5.2 *Given a geodesic $\gamma : I \subset \mathbb{R} \to \mathcal{M}$ and its tangent vector $\mathbf{t}_\gamma$, one has $\frac{d\mathbf{t}_\gamma}{ds} = \mathbf{0}$.*

Proof Recalling that $t^i = \frac{dx^i}{ds}$:

$$\frac{d\mathbf{t}_\gamma}{ds} = t^i_{;k}\frac{dx^k}{ds}\mathbf{e}_i = \left(\frac{\partial t^i}{\partial x^k} + \Gamma^i_{jk}t^j\right)\frac{dx^k}{ds}\mathbf{e}_i = \left(\frac{d^2x^i}{ds^2} + \Gamma^i_{jk}\frac{dx^j}{ds}\frac{dx^k}{ds}\right)\mathbf{e}_i = \mathbf{0}.$$

□

5.1.4 Electromagnetic and Gravitational Fields

In the case of a particle subject to an electromagnetic field on a curved manifold:

$$S = -m\int_{\tau_1}^{\tau_2}\sqrt{-g_{ij}dx^i dx^j} + q\int_{\tau_1}^{\tau_2} A_i\frac{dx^i}{d\tau}d\tau. \tag{5.19}$$

Carrying out the computations, one finds the equation of motion:

$$\frac{d^2x^r}{d\tau^2} + \Gamma^r_{ij}\frac{dx^i}{d\tau}\frac{dx^j}{d\tau} - \frac{q}{m}F^r{}_i\frac{dx^i}{d\tau} = 0 \tag{5.20}$$

which confirms the identification of the geometric term with the gravitational force.

5.1.5 Equivalence Principle

The geometric interpretation of free-fall motion is closely tied to the equivalence principle, which asserts the local indistinguishability between gravitational effects and accelerated reference frames.

In the presence of a gravitational field, i.e., on a curved manifold, it is possible to choose coordinates, called free-fall coordinates, such that, when restricted to a sufficiently small region of spacetime (so that curvature can be neglected locally), all physical laws have the same form: this is possible because, in that region, the metric tensor can be brought, by a change of reference frame, to the Minkowski metric, thereby recovering physics in flat spacetime.

For this to happen:

$$\eta_{ij} = \frac{\partial x^k}{\partial y^i}\frac{\partial x^l}{\partial y^j} g_{kl} \quad \Longrightarrow \quad \eta = \mathsf{D}^\mathsf{T} g \mathsf{D} \tag{5.21}$$

with $\mathsf{D} = \mathsf{J}^{-1}$. This is a similarity relation.

In such a reference frame, the equation of motion is that of the relativistic free particle:

$$\frac{d^2 y^i}{d\tau^2} = 0, \quad d\tau^2 = -\eta_{ij} dx^i dx^j. \tag{5.22}$$

These coordinates are also called free-fall or Fermi coordinates.

One can recover the geodesic equation from Eq. (5.22):

$$0 = \frac{d}{d\tau}\left(\frac{\partial y^i}{\partial x^k}\frac{dx^k}{d\tau}\right) = \frac{\partial y^i}{\partial x^k}\frac{d^2 x^k}{d\tau^2} + \frac{\partial^2 y^i}{\partial x^l \partial x^k}\frac{dx^k}{d\tau}\frac{dx^l}{d\tau}. \tag{5.23}$$

Multiplying by $\frac{\partial x^m}{\partial y^i}$:

$$\frac{d^2 x^m}{d\tau^2} + \frac{\partial x^m}{\partial y^i}\frac{\partial^2 y^i}{\partial x^l \partial x^k}\frac{dx^k}{d\tau}\frac{dx^l}{d\tau} \tag{5.24}$$

which corresponds to the geodesic equation derived earlier from the variational principle, if we identify:

$$\Gamma^m_{kl} = \frac{\partial x^m}{\partial y^i}\frac{\partial^2 y^i}{\partial x^l \partial x^k}. \tag{5.25}$$

This way of approximating the curved spacetime locally with the Minkowski spacetime is also known as Einstein's equivalence principle.

5.1.5.1 Newtonian Equivalence Principle

A principle of equivalence is also present in Newtonian physics. Consider a massive particle, described by the equation of motion:

$$m_\mathrm{i}\frac{d^2 \mathbf{x}}{dt^2} = m_\mathrm{g}\mathbf{g} + \sum_k \mathbf{F}_k. \tag{5.26}$$

In this case, the free-fall coordinates are defined by:

$$\mathbf{y} = \mathbf{x} - \frac{1}{2}\mathbf{g}t^2. \tag{5.27}$$

In this reference frame, the equation of motion becomes:

$$m_\mathrm{i}\left[\frac{d^2\mathbf{y}}{dt^2} + \mathbf{g}\right] = m_\mathrm{g}\mathbf{g} + \sum_k \mathbf{F}_k. \tag{5.28}$$

The Newtonian equivalence principle is an experimental fact and states that:

$$m_\mathrm{i} = m_\mathrm{g} \tag{5.29}$$

Thus, for a particle subject to no other forces, in free-fall coordinates the equation of motion becomes that for a free particle:

$$\frac{d^2\mathbf{y}}{dt^2} = \mathbf{0}. \tag{5.30}$$

It follows that free-fall coordinates "follow" the motion of a body in a gravitational field; similarly, Fermi coordinates identify a reference frame that follows the motion of the particle within curved spacetime.

5.1.6 Static Weak Gravitational Fields

Affine connections constitute a generalization of the Newtonian gravitational force field, whereas the metric tensor is a generalization of the Newtonian gravitational potential.

For simplicity, adopt the weak-field (spatial components negligible with respect to the temporal one) and static (time derivatives negligible) approximations: the geodesic equation becomes:

$$\frac{du^r}{d\tau} + \Gamma^r_{00}\left(u^0\right)^2 = 0. \tag{5.31}$$

Since, under the adopted approximation, $g_{k0,0} = g_{0k,0} = 0$, one has $\Gamma^r_{00} = -\frac{1}{2}g^{rk}g_{00,k}$. In the weak-field approximation one may write:

$$g_{ij} = \eta_{ij} + h_{ij}, \quad \left|h_{ij}\right| \ll 1. \tag{5.32}$$

Neglecting terms of order $O(h^2)$, one finds $\Gamma^r_{00} = -\frac{1}{2}h_{00,r}$, with $r = 1, 2, 3$. Using $u^0 = \frac{dx^0}{d\tau} \approx c$ in the non-relativistic limit, one finds

$$\frac{d^2x^r}{d\tau^2} = \frac{1}{2}\left(\frac{dx^0}{d\tau}\right)^2 h_{00,r} \approx \frac{c^2}{2}h_{00,r}.$$

Since $dt \approx d\tau$ in this limit, this becomes

$$\frac{d^2\mathbf{x}}{dt^2} = \frac{c^2}{2}\nabla h_{00}.$$

Comparing with the Newtonian equation

$$\mathbf{a} = -\nabla\Phi,$$

we obtain

$$h_{00} = -\frac{2\Phi}{c^2}. \tag{5.33}$$

Φ/c^2 is dimensionless and is $\sim 10^{-39}$ at the surface of a proton, $\sim 10^{-9}$ at Earth's surface, $\sim 10^{-6}$ at the Sun's surface, and $\sim 10^{-4}$ at the surface of a white dwarf.

5.1.6.1 Gravitational Time Dilation

Within the static weak-field approximation it is possible to estimate the effect of gravitational time dilation. For a clock, its proper time (time measured by the clock) is:

$$d\tau = \frac{1}{c}\sqrt{-g_{ij}dx^i dx^j} = \sqrt{-g_{00}}dt = \sqrt{1 - \frac{2GM}{c^2 r}}dt \tag{5.34}$$

where dt is the time measured in empty space. One sees that $d\tau < dt$.

Considering two bodies on Earth at distances from the center r and $r + h$:

$$\frac{T_2}{T_1} = \sqrt{\frac{1 - \frac{2GM}{c^2(r+h)}}{1 - \frac{2GM}{c^2 r}}} \approx 1 + \frac{gh}{c^2} \tag{5.35}$$

where $g \equiv \frac{GM}{r^2}$. From this follows the gravitational redshift (or blueshift), verified experimentally.

5.2 Riemann Tensor

While the equivalence principle allows gravity to be locally transformed away, it cannot eliminate tidal effects. These genuinely gravitational phenomena are captured by spacetime curvature, encoded in the Riemann tensor.

Definition 5.3 The Riemann tensor is defined by

$$R^i{}_{jkl} = \partial_k \Gamma^i_{lj} - \partial_l \Gamma^i_{kj} + \Gamma^i_{km}\Gamma^m_{lj} - \Gamma^i_{lm}\Gamma^m_{kj}. \tag{5.36}$$

The Riemann tensor can also be viewed as the commutator of covariant derivatives:

Proposition 5.3 $R^i_{mnk} = [\partial_k + \Gamma_k, \partial_n + \Gamma_n]^i{}_m$.

Proof Consider the affine connections as matrices $[\Gamma_k]^i{}_m = \Gamma^i_{km}$:

$$\begin{aligned}[\partial_k + \Gamma_k, \partial_n + \Gamma_n]^i{}_m &= \left[\Gamma_{n,k} - \Gamma_{k,n} + \Gamma_k\Gamma_n - \Gamma_n\Gamma_k\right]^i{}_m \\ &= \Gamma^i_{nm,k} - \Gamma^i_{km,n} + \Gamma^i_{kj}\Gamma^j_{nm} - \Gamma^i_{nj}\Gamma^j_{km}.\end{aligned}$$

□

Thus, equivalently, one can write:

$$[\nabla_k, \nabla_n]\, V^i = R^i_{mnk} V^m. \tag{5.37}$$

The physical meaning of the Riemann tensor is therefore that of performing the parallel transport of a vector around a loop: if there were no curvature, the operation would leave the vector unchanged; indeed, the Riemann tensor is also called the *curvature tensor*.

5.2.1 Geometric Meaning of the Riemann Tensor via Parallel Transport

Let (M, g) be a (pseudo-)Riemannian manifold with Levi–Civita connection ∇. The Riemann curvature tensor is defined by the commutator of covariant derivatives:

$$(\nabla_k \nabla_m - \nabla_m \nabla_k) V^i = R^i{}_{jkm}\, V^j, \tag{5.38}$$

for any vector V^i.

This has a direct geometric interpretation in terms of parallel transport around a tiny loop spanned by two infinitesimal vectors.

5.2.1.1 Infinitesimal Loop Spanned by Two Displacements

Let us pick a point $p \in M$, a tangent vector V^i at p, and two small displacement vectors A^k and B^m in the tangent space T_pM. Let us consider the following closed loop as depicted in Fig. 5.1:

1. move an infinitesimal amount along A,
2. then along B,
3. then back along $-A$,
4. then back along $-B$,

so that we return to p. Hence, we parallel–transport the vector V^i all the way around this loop. Our aim is to study the net change ΔV^i in V^i after coming back to p.

Transporting first along A and then along B can be written, to second order in the small parameters, as

$$V^i \mapsto V^i + A^k \nabla_k V^i + B^m \nabla_m V^i + B^m A^k \nabla_m \nabla_k V^i + O(A^2, B^2, AB^2, A^2B). \tag{5.39}$$

If we, instead, go first along B and then along A, we get

$$V^i \mapsto V^i + A^k \nabla_k V^i + B^m \nabla_m V^i + A^k B^m \nabla_k \nabla_m V^i + O(A^2, B^2, AB^2, A^2B). \tag{5.40}$$

Subtracting (5.40) from (5.39) isolates the failure of the two orders of transport to agree:

$$\Delta V^i = (\nabla_k \nabla_m - \nabla_m \nabla_k) V^i \; A^k B^m = R^i{}_{jkm} \, V^j \, A^k B^m. \tag{5.41}$$

Thus the curvature tensor $R^i{}_{jkm}$ measures exactly how much a vector fails to come back to itself after parallel transport around that infinitesimal parallelogram.

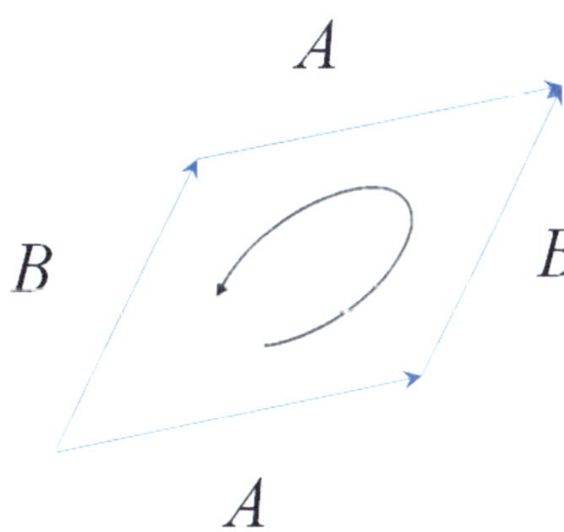

Fig. 5.1 Small loop made of two displacement vectors A and B lying on the manifold M along which a vector V^i (not shown) is parallel–transported

5.2.1.2 Geometric Meaning

If $R^i{}_{jkm} = 0$ at p, then $\Delta V^i = 0$ for any A, B, V: parallel transport around arbitrarily small loops is path-independent. The space is *locally flat* (no curvature) at that point. If $R^i{}_{jkm} \neq 0$, then the vector "twists" or "rotates" when carried around the loop. The amount of twisting is proportional to the area enclosed by the closed loop.

In other words, the Riemann tensor encodes the infinitesimal holonomy: it tells how much vectors are rotated/changed by parallel transport around infinitesimal closed loops.

Proposition 5.4 $R^i_{mnk} = -R^i_{mkn}$. *The Riemann tensor is antisymmetric with respect to exchanging the last two indices. This fact is due to the antisymmetry induced by the commutation bracket in the definition of the Riemann tensor.*

Definition 5.4 The *Ricci tensor* is the contraction $R_{mk} \equiv R^n_{mnk}$.

Definition 5.5 The *Ricci scalar* is the contraction $R \equiv g^{mk} R_{mk}$.

5.2.2 *Example: Curvature of the 2-Sphere*

It is a useful exercise to compute the curvature of a sphere.

Consider a 2-sphere of radius r with coordinates $(\theta, \phi) \in (0, \pi) \times [0, 2\pi)$: the possible Levi-Civita connections are 8, but only 6 are independent due to symmetry in the lower indices. Taking $\mathbb{R}^3$ as the embedding space:

$$\mathbf{p} = r\,(\sin\theta\cos\phi, \sin\theta\sin\phi, \cos\theta)\,.$$

Computing the basis vectors:

$$\mathbf{e}_\theta = \frac{\partial \mathbf{p}}{\partial \theta} = r\,(\cos\theta\cos\phi, \cos\theta\sin\phi, -\sin\theta) \equiv r\hat{\mathbf{e}}_\theta$$

$$\mathbf{e}_\phi = \frac{\partial \mathbf{p}}{\partial \phi} = r\sin\theta\,(-\sin\phi, \cos\phi, 0) \equiv r\sin\theta\hat{\mathbf{e}}_\phi.$$

Thus one finds the metric tensor of the sphere:

$$g_{ij} = \begin{bmatrix} r^2 & 0 \\ 0 & r^2\sin^2\theta \end{bmatrix} \qquad g^{ij} = \begin{bmatrix} r^{-2} & 0 \\ 0 & r^{-2}\sin^{-2}\theta \end{bmatrix}.$$

The dual basis vectors are therefore $\mathbf{e}^\theta = r^{-1}\hat{\mathbf{e}}_\theta$ and $\mathbf{e}^\phi = r^{-1}\sin^{-1}\theta\hat{\mathbf{e}}_\phi$. The derivatives of the basis vectors are:

$$\mathbf{e}_{\theta,\theta} = -r\hat{\mathbf{e}}_r \qquad \mathbf{e}_{\theta,\phi} = \mathbf{e}_{\phi,\theta} = r\cos\theta\hat{\mathbf{e}}_\phi \qquad \mathbf{e}_{\phi,\phi} = -r\sin\theta\,(\cos\phi,\sin\phi,0)\,.$$

Recalling that $\Gamma^i_{\ jk} \equiv \mathbf{e}^i \cdot \mathbf{e}_{k,j}$ and that $[\Gamma_k]^i_{\ m} = \Gamma^i_{km}$, one finds:

$$\Gamma_\theta = \begin{bmatrix} 0 & 0 \\ 0 & \cot\theta \end{bmatrix}, \qquad \Gamma_\phi = \begin{bmatrix} 0 & -\sin\theta\cos\theta \\ \cot\theta & 0 \end{bmatrix} \tag{5.42}$$

$$\Longrightarrow \quad \left[\Gamma_\theta, \Gamma_\phi\right] = -\left[\Gamma_\phi, \Gamma_\theta\right] = \begin{bmatrix} 0 & \cos^2\theta \\ \cot^2\theta & 0 \end{bmatrix} \tag{5.43}$$

For computing the Ricci tensor only 4 components of the Riemann tensor are needed:

$$R^\theta_{\ \theta\theta\theta} = [\partial_\theta + \Gamma_\theta, \partial_\theta + \Gamma_\theta]^\theta_{\ \theta} = 0$$

$$R^\phi_{\ \theta\phi\theta} = \left[\partial_\theta + \Gamma_\theta, \partial_\phi + \Gamma_\phi\right]^\phi_{\ \theta} = [\Gamma_{\phi,\theta}]^\phi_{\ \theta} + \left[\Gamma_\theta, \Gamma_\phi\right]^\phi_{\ \theta} = (\cot\theta)_{,\theta} + \cot^2\theta = -1$$

$$R^\theta_{\ \phi\theta\phi} = \left[\partial_\phi + \Gamma_\phi, \partial_\theta + \Gamma_\theta\right]^\theta_{\ \phi} = -[\Gamma_{\phi,\theta}]^\theta_{\ \phi} = \cos^2\theta - \sin^2\theta - \cos^2\theta = -\sin^2\theta$$

$$R^\phi_{\ \phi\phi\phi} = \left[\partial_\phi + \Gamma_\phi, \partial_\phi + \Gamma_\phi\right]^\phi_{\ \phi} = 0.$$

The non-vanishing components of the Ricci tensor are:

$$R_{\theta\theta} = R^\theta_{\ \theta\theta\theta} + R^\phi_{\ \theta\phi\theta} = -1$$

$$R_{\phi\phi} = R^\theta_{\ \phi\theta\phi} + R^\phi_{\ \phi\phi\phi} = \quad \sin^2\theta.$$

Since $R = g^{mk}R_{mk} = g^{\theta\theta}R_{\theta\theta} + g^{\phi\phi}R_{\phi\phi}$, one finds the scalar curvature of a 2-sphere:

$$R = \frac{2}{r^2}.$$

This agrees with the known result that the Gaussian curvature is

$$K = \frac{1}{r^2},$$

and in two dimensions one has

$$R = 2K.$$

This confirms the Gauss–Bonnet theorem, which relates the curvature K to the scalar curvature: for a 2-sphere $K = \frac{1}{r^2}$, hence $K = -\frac{R}{2}$ (as stated by the theorem).

5.3 Einstein's Field Equations

Having characterized how curvature affects the motion of matter, we now address the inverse problem: how matter and energy determine the curvature of spacetime itself.

An isolated particle in curved spacetime moves according to the geodesic equation. In the presence of a matter field, however, the situation becomes a bit more complicated because the curvature of the spacetime is induced by the properties of the matter field.

A matter field, in which there are flows of matter and energy, is described by the energy–momentum tensor T_{ij}, defined so as to satisfy energy conservation $\nabla_i T^i{}_j = 0$.

For example, for an isotropic ideal fluid at rest with density ρ and pressure p, the energy–momentum tensor is $T_{ij} = pg_{ij} + (p + \rho)u_i u_j$, with $u^i = \frac{dx^i}{d\tau} = \gamma(c, \mathbf{v})$.

The link between the matter field and the curvature of spacetime is given by the *Einstein field equation*:

$$R_{ij} - \frac{1}{2}g_{ij}R = \frac{8\pi G}{c^4}T_{ij}. \tag{5.44}$$

Contracting with g^{ij}, since $g^{ij}g_{ij} = \delta^i_i = 4$, and defining the trace of the energy–momentum tensor $T \equiv T^i_i$, one finds $R = \frac{8\pi G}{c^4}T$, hence an equivalent form of the field equation is:

$$R_{ij} = \frac{8\pi G}{c^4}\left(T_{ij} - \frac{1}{2}Tg_{ij}\right). \tag{5.45}$$

In vacuum and on small scales (isotropic space), neglecting dark energy, the field equation becomes $R_{ij} = 0$.

5.3.1 *Einstein–Hilbert Action*

Just as particle dynamics follow from an action principle, the dynamics of spacetime geometry itself can be obtained by varying an appropriate gravitational action.

It is possible to construct an action from which the field equation follows: being a scalar, it is invariant under arbitrary coordinate transformations, hence the tensor equations derived from it are invariant under such transformations.

To determine a suitable action, one needs a scalar depending on the metric: the Ricci scalar. Recalling that the 4-form volume element is $\sqrt{g}\,d^4x$, one obtains the *Einstein–Hilbert action* in vacuum:

$$S_{\text{EH}} = -\frac{c^4}{16\pi G}\int *R = -\frac{c^4}{16\pi G}\int R\sqrt{g}\,d^4x. \tag{5.46}$$

Proposition 5.5 *Einstein's field equation follows from the Einstein–Hilbert action.*

Proof In the absence of matter fields, given a variation of the metric δg_{ij} that vanishes at infinity, the first-order variation of the action can be written as:

$$\delta S_{\text{EH}} = -\frac{c^4}{16\pi G}\int \delta(R\sqrt{g})\,d^4x = -\frac{c^4}{16\pi G}\int\left(\sqrt{g}\frac{\delta R}{\delta g^{ij}} + R\frac{\delta\sqrt{g}}{\delta g^{ij}}\right)\delta g^{ij}\,d^4x$$

For the first term, using $\delta R^i_{klm} = \nabla_l(\delta\Gamma^i_{mk}) - \nabla_m(\delta\Gamma^i_{lk})$:

$$\delta R = R_{ij}\delta g^{ij} + g^{ij}\delta R_{ij} = R_{ij}\delta g^{ij} + g^{ij}\delta R^k_{ikj} \tag{5.47}$$

$$= R_{ij}\delta g^{ij} + g^{ij}\left[\nabla_k(\delta\Gamma^k_{ji}) - \nabla_j(\delta\Gamma^k_{ki})\right] \tag{5.48}$$

$$= R_{ij}\delta g^{ij} \tag{5.49}$$

where the vanishing of the last term was shown by York [1] Gibbons and Hawking [2], in the 1970s. As for the second term, recall Jacobi's formula $\frac{\delta \det \mathsf{A}}{\delta A_{ij}} = (A^{-1})_{ij}\det \mathsf{A}$:

$$\delta\sqrt{g} = \frac{1}{2\sqrt{g}}\delta g = \frac{1}{2\sqrt{g}}g g^{ij}\delta g_{ij} = -\frac{1}{2}\sqrt{g}g_{ij}\delta g^{ij}$$

where in the last equality we used $g^{ij}\delta g_{ij} = -g_{ij}\delta g^{ij}$, which is a general property:

$$k^{-1}\frac{\delta k}{\delta p} + \frac{\delta k^{-1}}{\delta p}k = \frac{\delta(k^{-1}k)}{\delta p} = \frac{\delta 1}{\delta p} = 0 \quad\Longrightarrow\quad \delta g^{ij} = -g^{il}(\delta g_{lm})g^{mj}$$

Consequently:

$$\delta S_{\text{EH}} = -\frac{c^4}{16\pi G}\int\left(R_{ij} - \frac{1}{2}g_{ij}R\right)\delta g^{ij}\sqrt{g}\,d^4x$$

hence:

$$\delta S_{\text{EH}} = 0 \quad\Longleftrightarrow\quad R_{ij} - \frac{1}{2}g_{ij}R = 0$$

In the presence of a matter field, one adds a term of the form:

$$\delta S_m = -\frac{1}{2}\int T_{ij}\delta g^{ij}\sqrt{g}\,d^4x \quad \Longleftrightarrow \quad T_{ij} = -2\frac{1}{\sqrt{g}}\frac{\delta S_m}{\delta g^{ij}}$$

This form can be justified as follows.

Consider a (classical) matter field or collection of fields ψ on a spacetime with metric g_{ij}. The matter action is assumed to be of the form

$$S_m[g,\psi] = \int d^4x\,\sqrt{g}\,\mathcal{L}_m(\psi,\nabla\psi,g^{ij}), \tag{5.50}$$

where $g = \det(g_{ij})$ and $\mathcal{L}_m$ is the matter Lagrangian density. We assume *minimal coupling*: $\mathcal{L}_m$ depends on the metric only through contractions like $g^{ij}\partial_\mu\psi\,\partial_\nu\psi$ etc., and not on derivatives of g_{ij}.

The stress–energy tensor T_{ij} is defined as the functional derivative of the matter action with respect to the inverse metric g^{ij}:

$$T_{ij} \equiv -\frac{2}{\sqrt{g}}\frac{\delta S_m}{\delta g^{ij}} \tag{5.51}$$

(with ψ held fixed when taking $\delta/\delta g^{ij}$). We now show how this formula arises.

Vary $g^{ij} \mapsto g^{ij} + \delta g^{ij}$ while keeping ψ fixed. We have

$$\delta S_m = \int d^4x\left[\delta(\sqrt{g})\,\mathcal{L}_m + \sqrt{g}\,\delta\mathcal{L}_m\right]. \tag{5.52}$$

First, the standard identity for the variation of $\sqrt{g}$ is

$$\delta\sqrt{g} = -\tfrac{1}{2}\sqrt{g}\,g_{ij}\,\delta g^{ij}. \tag{5.53}$$

This follows from $\delta\ln g = g_{ij}\,\delta g^{ij}$ and $\sqrt{g} = g^{1/2}$.

Second, because $\mathcal{L}_m$ depends on the metric but not on its derivatives, its variation at fixed ψ is

$$\delta\mathcal{L}_m = \frac{\partial\mathcal{L}_m}{\partial g^{ij}}\,\delta g^{ij}. \tag{5.54}$$

Insert (5.53) and (5.54) into (5.52):

$$\delta S_m = \int d^4x\left[\left(-\tfrac{1}{2}\sqrt{g}\,g_{ij}\,\delta g^{ij}\right)\mathcal{L}_m + \sqrt{g}\,\frac{\partial\mathcal{L}_m}{\partial g^{ij}}\,\delta g^{ij}\right] \tag{5.55}$$

$$= \frac{1}{2}\int d^4x\,\sqrt{g}\left[-g_{ij}\,\mathcal{L}_m + 2\,\frac{\partial\mathcal{L}_m}{\partial g^{ij}}\right]\delta g^{ij}. \tag{5.56}$$

By *definition* of the stress–energy tensor, we also want to be able to write any metric variation as

$$\delta S_m = \frac{1}{2} \int d^4x \sqrt{g}\, T_{ij}\, \delta g^{ij}. \tag{5.57}$$

Because δg^{ij} is arbitrary, we can match the integrands of (5.56) and (5.57) term by term, giving

$$T_{ij} = -2 \frac{1}{\sqrt{g}} \frac{\delta S_m}{\delta g^{ij}} = -2 \frac{\partial \mathcal{L}_m}{\partial g^{ij}} + g_{ij}\, \mathcal{L}_m. \tag{5.58}$$

This is exactly (5.51). The two expressions in (5.58) are equivalent because of how δS_m was computed.

Comments

- T_{ij} defined this way is symmetric (for minimally coupled matter). This is the stress–energy tensor that appears in Einstein's equations.
- If the matter fields satisfy their own Euler–Lagrange equations, and if S_m is diffeomorphism-invariant, then $\nabla^\mu T_{ij} = 0$ (covariant conservation).
- Physically, T_{ij} encodes local energy density, momentum density, and stresses (pressures and shear). In a local rest frame of a perfect fluid with 4-velocity u^μ, one finds

  $$T_{ij} = (\rho + p)\, u_\mu u_\nu + p\, g_{ij} \quad (c = 1),$$

  where ρ is energy density and p is pressure.

Putting things together, we have thus obtained the Einstein field equation in the presence of matter and energy fields:

$$-\frac{c^4}{16\pi G} \left(R_{ij} - \frac{1}{2} g_{ij} R \right) - \frac{1}{2} T_{ij} = 0,$$

which provides the fundamental link between matter fields and the curvature of spacetime. The simplest analytical solution to the Einstein field equation is the one found by Schwarzschild for a static, spherically symmetric vacuum surrounding a spherical massive object. The solution provides an explicit form for the components of the metric tensor g_{ij} in spherical coordinates [3].

General relativity thus emerges as a unified geometric theory in which both the motion of matter and the dynamics of spacetime arise from variational principles.

References

1. J.W. York, Phys. Rev. Lett. **28**, 1082 (1972). https://doi.org/10.1103/PhysRevLett.28.1082. https://link.aps.org/doi/10.1103/PhysRevLett.28.1082
2. G.W. Gibbons, S.W. Hawking, Phys. Rev. D **15**, 2752 (1977). https://doi.org/10.1103/PhysRevD.15.2752. https://link.aps.org/doi/10.1103/PhysRevD.15.2752
3. L.D. Landau, E.M. Lifshitz, *The Classical Theory of Fields*, Course of Theoretical Physics, vol. 2, 4th edn. (Butterworth-Heinemann, Oxford, 1975)

Chapter 6
Manifolds and Diffeomorphisms

Abstract In this chapter we formalize the basic notions of differentiable manifolds and of diffeomorphisms that we have used rather informally in the previous chapters. This also prompts us to provide some connections between topology and differential geometry which are useful in the physical sciences. This chapter completes the geometric framework underlying the previous chapters by making precise the notions of locality, smooth structure, and topology that are essential for understanding curvature, gauge fields, and topological invariants in physics. Several worked-out examples concerning topics of current research in physics, ranging from quantum geometry to topological defects to elasticity theory, will be presented.

6.1 Differentiable Manifolds

The geometric structures introduced in the previous chapters—covariant derivatives, curvature, and geodesic motion—implicitly rely on the notion of a smooth manifold. We now formalize this concept and make precise the assumptions that were previously used at an intuitive level.

Informally, a manifold is an object that on a large scale can have non-trivial curvature properties, but that locally appears as a Euclidean (flat) space.

One immediately sees a similarity with Einstein's equivalence principle: physical laws reduce to those of special relativity in sufficiently small neighborhoods of spacetime; since gravity emerges from the curvature properties of spacetime, this is equivalent to saying that spacetime is described by a differentiable manifold.

Mathematically, this local flatness is encoded in the existence of coordinate charts that identify small neighborhoods of the manifold with open subsets of $\mathbb{R}^n$.

Definition 6.1 A *Euclidean space* is defined as $\mathbb{R}^n \equiv \{(x^1, \ldots, x^n) : x^j \in \mathbb{R}\}$ endowed with the positive definite flat metric $g_{ij} = \delta_{ij}$, where δ_{ij} is the Kronecker delta.

The local identification of a manifold with $\mathbb{R}^n$ is not an equality of metrics, but allows one to equivalently define the notions of functions, coordinates, operators, etc.

A. Zaccone, *Differential Geometry and Group Theory*,
https://doi.org/10.1007/978-3-032-27518-9_6

6.1.1 Examples of Manifolds

Examples of manifolds are the n-sphere S^n, namely the locus of points in $\mathbb{R}^{n+1}$ at unit distance from the origin, and the n-torus T^n, generated by identifying the opposite sides of an n-cube. This process is schematically depicted in Fig. 6.1 for the case of the T^2 torus.

A *Riemann surface* is defined as an orientable, compact, boundaryless, 2-dimensional differentiable manifold; Riemann surfaces can be characterized by their genus g, that is the number of holes of such manifold: a necessary condition for two Riemann surfaces to be homotopic is that they have the same genus.

The group of continuous rotations in $\mathbb{R}^n$ forms a manifold: in general, *Lie groups* are manifolds endowed with a group structure; for example $\mathrm{SO}(2) \cong S^1$.

6.1.2 Charts and Atlases

To make the idea of local Euclidean behavior precise, we must describe how local coordinate systems are defined, and how they are consistently related to one another across the manifold.

The manifold as a whole is constructed by smoothly attaching the various locally flat regions that compose it. Each of these locally flat regions are thus mapped onto an open set in the corresponding $\mathbb{R}^n$ space. We first need to clarify the concept of maps and of open sets.

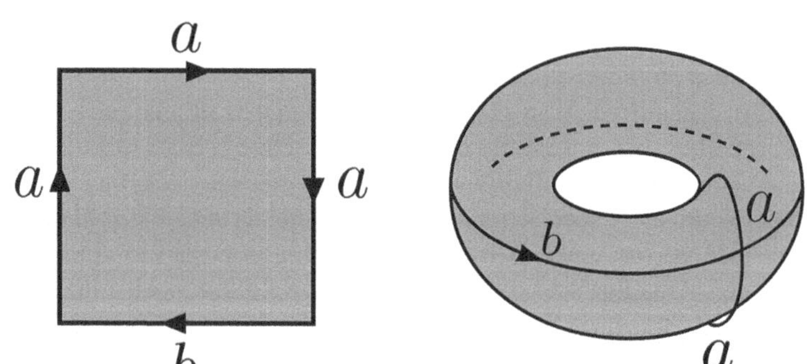

Fig. 6.1 A 2-torus, T^2, is generated from a square (i.e. a 2-cube) by identifying the opposite edges of the square. The square has opposite sides labeled (a) and (b). First, gluing the sides (a) together turns the square into a cylinder. Then, gluing the sides (b) of the cylinder together produces the torus

6.1.2.1 Maps

Definition 6.2 Given two sets M, N, a *map* (or function) $\varphi : M \to N$ is a rule that assigns to each element of M exactly one element of N. M is called the *domain* and $\varphi(M) \subseteq N$ the *image*.

Definition 6.3 A map $\varphi : M \to N$ is called *injective* if to each element of the image corresponds only one element of the domain.

Definition 6.4 A map $\varphi : M \to N$ is called *surjective* if $\varphi(M) = N$.

Definition 6.5 A map $\varphi : M \to N$ is called *bijective* if it is injective and surjective.

Definition 6.6 Given a bijective map $\varphi : M \to N$, one defines the inverse map $\varphi^{-1} : N \to M$ such that $\varphi \circ \varphi^{-1} = \mathrm{id}_N$ and $\varphi^{-1} \circ \varphi = \mathrm{id}_M$.

Recall that a map between Euclidean spaces $\varphi : \mathbb{R}^m \to \mathbb{R}^n$ is defined as:

$$\varphi : (x^1, \dots, x^m) \mapsto (\varphi^1(x^1, \dots, x^m), \dots, \varphi^n(x^1, \dots, x^m)) \tag{6.1}$$

which corresponds to a set of equations:

$$y^1 = \varphi^1(x^1, x^2, \dots, x^m) \tag{6.2}$$

$$y^2 = \varphi^2(x^1, x^2, \dots, x^m) \tag{6.3}$$

$$\vdots$$

$$y^n = \varphi^n(x^1, x^2, \dots, x^m) \tag{6.4}$$

These are functions in C^p if all partial derivatives up to order p exist and are continuous, and the whole map $\varphi : \mathbb{R}^m \to \mathbb{R}^n$ is C^p if each component function φ^i is C^p.

Definition 6.7 Given a map $\varphi : \mathbb{R}^m \to \mathbb{R}^n$, it is said to be of *class* C^p, $p \geq 0$, if all partial derivatives of φ^i up to order p exist and are continuous.

Definition 6.8 A map $\varphi \in C^\infty(\mathbb{R}^m, \mathbb{R}^n)$ is called *smooth*.

6.1.2.2 Diffeomorphisms Between Sets

In differential geometry, not all maps are equally meaningful: the physically relevant ones are those that preserve the smooth structure. This motivates the definition of diffeomorphisms.

Definition 6.9 Two sets M, N are called diffeomorphic, written $M \cong N$, if there exists a diffeomorphism between them, namely a bijective map $\varphi \in C^\infty(M, N)$ with inverse $\varphi^{-1} \in C^\infty(N, M)$.

The notion of diffeomorphism allows us to establish whether two spaces represent the same manifold: a diffeomorphism is an isomorphism between smooth manifolds. A simple intuitive example is provided by the isomorphism between the group SO(2) of all rotations in the 2D plane and the manifold S^1:

$$\mathrm{SO}(2) \cong S^1. \tag{6.5}$$

6.1.2.3 Open Balls and Open Sets

Definition 6.10 Given $\mathbf{x} \in \mathbb{R}^n$ and $r \in \mathbb{R}^+$, the *open ball* centered at $\mathbf{x}$ with radius r is defined as the set $\mathcal{B}_r(\mathbf{x}) \equiv \{\mathbf{y} \in \mathbb{R}^n : |\mathbf{y} - \mathbf{x}| < r\}$, with the norm determined by the metric.

The concept of open ball is illustrated in Fig. 6.2.

Definition 6.11 $V \subset \mathbb{R}^n$ is said to be *open* if $\forall \mathbf{x} \in V \, \exists r \in \mathbb{R}^+ : \mathcal{B}_r(\mathbf{x}) \subset V$.

An open set can thus be seen as a union of open balls, or as the interior of one or more $(n-1)$-dimensional surfaces.

Definition 6.12 A *chart* on a set $\mathcal{M}$ consists of a set $U \subset \mathcal{M}$ and a map $\varphi : U \to \mathbb{R}^n$ such that $\varphi(U) \subset \mathbb{R}^n$ is open.

The concept of a map between a sub-manifold U of$M$ and an open set of $\mathbb{R}^n$ is schematically illustrated in Fig. 6.3.

Definition 6.13 An *atlas* on a set $\mathcal{M}$ is a set of charts $\mathcal{A} = \{U_\alpha, \varphi_\alpha\}$ such that their union covers $\mathcal{M}$ and they are smoothly compatible: if $U_\alpha \cap U_\beta \neq \emptyset$, then the maps $\varphi_\alpha \circ \varphi_\beta^{-1}$ and $\varphi_\beta \circ \varphi_\alpha^{-1}$ are diffeomorphisms.

Transition maps are nothing but coordinate changes on the intersections $U_\alpha \cap U_\beta$.

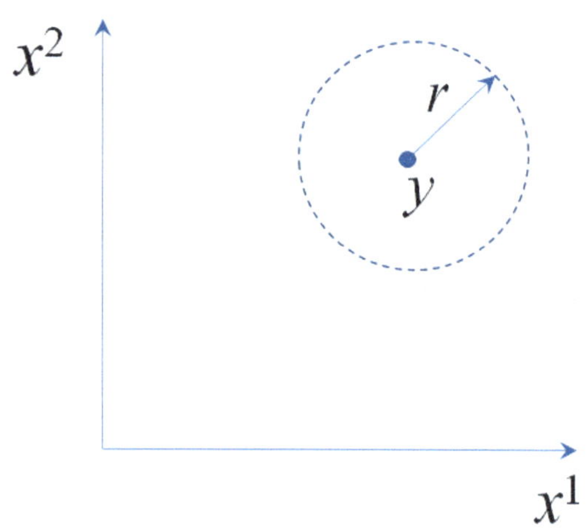

Fig. 6.2 Schematic depiction of an open ball in $\mathbb{R}^2$

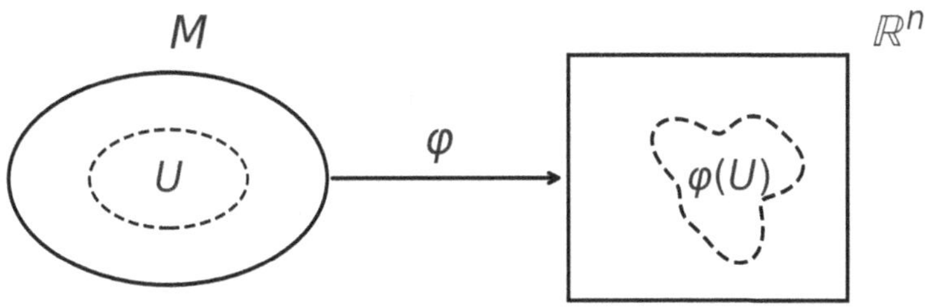

Fig. 6.3 A map φ sends points of a sub-manifold U of manifold M into points of an open set contained in $\mathbb{R}^n$. The map φ and the subset U, taken together, form a chart

Definition 6.14 A *differentiable manifold* is a set $\mathcal{M}$ with a maximal atlas $\mathcal{A}$ on it.

While differentiable structure is defined locally through charts, global properties of manifolds—such as whether a single chart suffices—are inherently topological.

Diffeomorphism is about the existence of a smooth bijective map with smooth inverse; that does not require maximality. Using maximal atlases ensures that two atlases which define the same smooth structure are not regarded as different structures on the manifold. The maximal atlas collects all charts compatible with that structure, making it unique.

Theorem 6.1 (Whitney Embedding Theorem) *Any smooth n-dimensional manifold can be smoothly embedded in* $\mathbb{R}^{2n+1}$.

Whitney's theorem is motivated by some topologically non-trivial surfaces, which cannot be embedded within a 3D space. The most famous example is, perhaps, the Klein bottle. This is a 2-dimensional surface which cannot be embedded in $\mathbb{R}^3$, but it can in $\mathbb{R}^4$.

6.1.3 *Covering of Manifolds with Charts*

One must note that most manifolds cannot be covered by a single chart.

Consider S^1 with the usual choice of coordinate $\varphi : p \in S^1 \mapsto \theta \in [0, 2\pi] \subset \mathbb{R}$: if 0 or 2π are included, then $\varphi(S^1)$ is not open in $\mathbb{R}$, thus two charts are necessary, as shown in Fig. 6.4.

Now let us consider S^2 with stereographic (or Mercator) projection: since the projection of the sphere $(x^1)^2+(x^2)^2+(x^3)^2 = 1$ onto $x^3 = -1$ is defined except at the north pole, two charts are necessary: the stereographic projection onto $x^3 = -1$, which excludes the north pole, and that onto $x^3 = 1$, which excludes the south pole. The former is schematically depicted in Fig. 6.5. We adopt the convention of projecting the sphere from the north pole onto the plane $x^3 = -1$ (rather than onto the equatorial plane $x^3 = 0$ as is often done in textbooks). This choice only changes the overall scale of the coordinates and simplifies some expressions below.

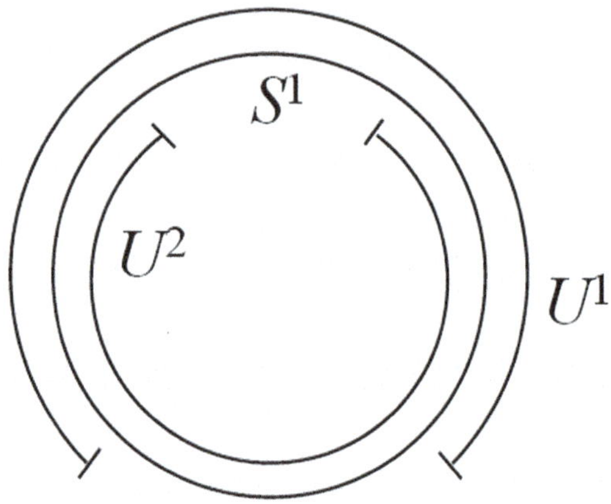

Fig. 6.4 The manifold S^1 requires two charts, U^1 and U^2 to be covered

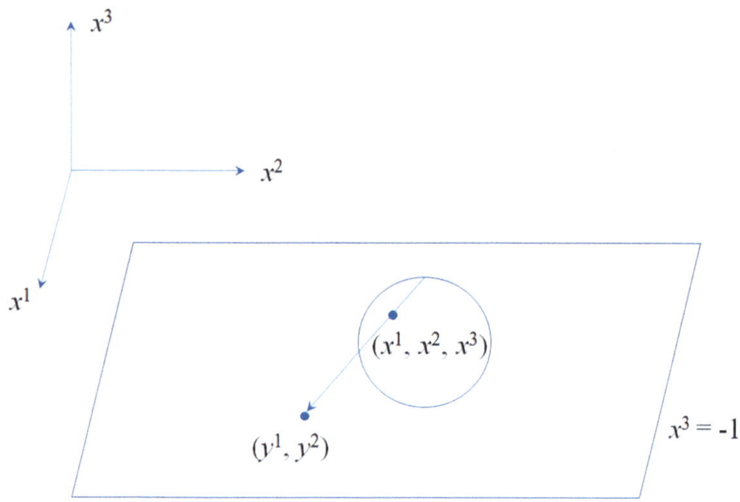

Fig. 6.5 The stereographic projection from the north pole covers the whole spherical surface of S^2 but leaves the north pole out, hence it does fully cover S^2

Defining y^1, y^2 and z^1, z^2 the coordinates on the considered planes, we find:

$$\varphi_{-1}(x^1, x^2, x^3) = \left(\frac{2x^1}{1 - x^3}, \frac{2x^2}{1 - x^3}\right) \equiv (y^1, y^2)$$

$$\varphi_1(x^1, x^2, x^3) = \left(\frac{2x^1}{1 + x^3}, \frac{2x^2}{1 + x^3}\right) \equiv (z^1, z^2)$$

These relations can be easily found by considering similar triangles or using a Moebius transformation, or analytically by considering the parametric form of a straight line which intersects the plane and the spherical surface.

The two charts intersect on $-1 < x^3 < 1$ and the transition map is:

$$\varphi_1 \circ \varphi_{-1}^{-1}(y^1, y^2) = \left(\frac{4y^1}{(y^1)^2 + (y^2)^2}, \frac{4y^2}{(y^1)^2 + (y^2)^2} \right) \equiv (z^1, z^2).$$

6.2 Tangent Space Revisited

Having defined manifolds as spaces that are locally Euclidean but globally curved, we now turn to the linear structures that capture local directions and variations on the manifold.

Definition 6.15 Given a manifold $\mathcal{M}$ and a point $p \in \mathcal{M}$, the *tangent space* to $\mathcal{M}$ at p is defined as the space $T_p\mathcal{M}$ of all tangent vectors to the manifold at that point.

It is possible to construct $T_p\mathcal{M}$ equivalently: consider a curve passing through p, that is a map $\varphi : I \subset \mathbb{R} \to \mathcal{M} : \lambda \mapsto \varphi(\lambda)$ such that $\exists \lambda_0 \in I : \varphi(\lambda_0) = p$. If the embedding space of $\mathcal{M}$ is $\mathbb{R}^n$, then at each point of the image the curve determines a vector $\tau_\varphi(\lambda)$ with components $\frac{dx^k}{d\lambda}(\lambda)$, called tangent vector: the map $p \mapsto \tau_\varphi(\lambda_0)$ depends on the chosen chart.

Defining $\mathcal{F}$ as the space of smooth functions on $\mathcal{M}$, each curve φ passing through p defines a vector $\frac{df}{d\lambda}$ for each $f \in \mathcal{F}$, called directional derivative: the map $f \mapsto \nabla_\varphi f(\lambda_0)$ does not depend on the chosen chart.

The tangent space $T_p\mathcal{M}$ can be seen as the space of directional derivative operators along curves passing through p.

Definition 6.16 Given a manifold $\mathcal{M}$ and a point $p \in \mathcal{M}$, the *cotangent space* to $\mathcal{M}$ at p is defined as the dual space $T_p^*\mathcal{M}$ of $T_p\mathcal{M}$.

6.2.1 *Tangent Space and Cotangent Space (Geometric Intuition)*

We work on a smooth manifold M, and fix a point $p \in M$.

According to our intuition, the tangent space T_pM is the set of all possible "velocity vectors" you could have if you pass through p. If you stand on a surface and start walking in some direction, your initial velocity vector lies in T_pM.

Formally, T_pM is a vector space consisting of tangent vectors at p. In local coordinates $(x^1, \ldots, x^n)$ around p, a natural basis is

$$\left\{ \left.\frac{\partial}{\partial x^1}\right|_p, \left.\frac{\partial}{\partial x^2}\right|_p, \ldots, \left.\frac{\partial}{\partial x^n}\right|_p \right\}.$$

Each basis vector $\frac{\partial}{\partial x^i}\Big|_p$ means: "change only the coordinate x^i at unit rate; keep the others fixed."

Definition 6.17

$$T_p^*M := \{\omega : T_pM \to \mathbb{R} \text{ linear}\}.$$

That is, T_p^*M is the **dual space** to T_pM: each element of T_p^*M (called a *covector* or *1-form*) is a linear "measurement rule" that takes a tangent vector and outputs a real number.

According to our intuition: a tangent vector is something you *do*: a direction plus a speed of motion through p. A cotangent vector is something you *use to measure*: it looks at a given direction of motion and tells you "how much in my direction are you moving?"

So if $v \in T_pM$ is a direction of motion and $\omega \in T_p^*M$ is a covector, then $\omega(v)$ is a real number. The map

$$T_p^*M \times T_pM \to \mathbb{R}, \qquad (\omega, v) \mapsto \omega(v)$$

is called the dual pairing.

In the same local coordinates $(x^1, \ldots, x^n)$, we can define the differentials

$$dx^1\Big|_p, \; dx^2\Big|_p, \; \ldots, \; dx^n\Big|_p.$$

Each $dx^i\Big|_p$ is an element of T_p^*M. It acts on a tangent vector by telling you how fast the x^i coordinate changes in that direction.

Concretely,

$$dx^i\Big|_p \left(\frac{\partial}{\partial x^j}\Big|_p \right) = \delta^i{}_j.$$

This is the defining property of a **dual basis**:

- $\left\{ \frac{\partial}{\partial x^j}\Big|_p \right\}$ is a basis of T_pM.
- $\left\{ dx^i\Big|_p \right\}$ is the uniquely corresponding dual basis of T_p^*M.

The interpretation is as follows:

- $\frac{\partial}{\partial x^j}\Big|_p$: "move in the x^j direction with unit speed."
- $dx^i\Big|_p$: "tell me the rate of change of the coordinate x^i along your motion."
- The pairing gives 1 if they match, 0 if they don't.

Let $f : M \to \mathbb{R}$ be a smooth scalar function. Its differential at p, usually denoted $df\big|_p$, is defined by

$$df\big|_p(v) := v[f],$$

i.e. "the directional derivative of f in the direction v," evaluated at p.

One should always remember that: $df\big|_p$ is linear in v. Therefore $df\big|_p$ is an element of T_p^*M.

So gradients/differentials live in the cotangent space. Tangent vectors tell you *which way you're going*; cotangent vectors (like df) tell you *how fast something changes if you go that way*.

In coordinates, if

$$f = f(x^1, \ldots, x^n),$$

then

$$df\Big|_p = \sum_{i=1}^{n} \frac{\partial f}{\partial x^i}\Big|_p \, dx^i\Big|_p.$$

If $v = v^j \frac{\partial}{\partial x^j}\big|_p$, then

$$df\big|_p(v) = \sum_{i=1}^{n} \frac{\partial f}{\partial x^i}\Big|_p \, dx^i\Big|_p \left(v^j \frac{\partial}{\partial x^j}\Big|_p \right) = \sum_{i=1}^{n} \frac{\partial f}{\partial x^i}\Big|_p \, v^i.$$

That is the usual directional derivative.

In general relativity or field theory:

- A tangent vector at an event can be the 4-velocity u^μ of a particle.
- A covector can be the gradient of a scalar field, $\partial_\mu \phi$.

Then the pairing

$$u^\mu \, \partial_\mu \phi$$

tells you how fast the field ϕ changes along the worldline of the particle. Here:

- u^μ is in T_pM (a direction in spacetime),
- $\partial_\mu \phi$ is in T_p^*M (a covector field),
- their contraction is a number (a physical rate of change).

This is exactly the abstract definition:

$$\text{(cotangent vector)(tangent vector)} = \text{real number}.$$

To summarize: T_pM (tangent space) is the vector space of all possible directions/velocities through p. T_p^*M (cotangent space) is the vector space of all linear functionals on T_pM. Elements of T_p^*M are also called covectors. In coordinates, $\left\{\frac{\partial}{\partial x^i}\Big|_p\right\}$ is a basis of T_pM, and $\left\{dx^i\big|_p\right\}$ is its dual basis in T_p^*M, satisfying

$$dx^i\Big|_p\left(\frac{\partial}{\partial x^j}\Big|_p\right) = \delta^i{}_j.$$

Differentials of scalar fields (like df) are natural examples of cotangent vectors: they eat a direction and tell you the rate of change of f in that direction.

This construction underlies the definition of differential forms introduced in Chap. 3, which are fields of antisymmetric covariant tensors built from the cotangent spaces.

6.3 Topological Defects

Beyond local differential structure, manifolds and fields defined on them may possess global properties that cannot be detected by local measurements alone. These properties are topological in nature, and often manifest physically as defects or "quantized" responses.

In crystals, whenever an atomic sheet is displaced with respect to the ideal ordered atomic configuration of the lattice, we are in presence of a dislocation. Dislocations in crystals are examples of *topological line defects*. In a continuum description, by "topological defect" one means a localized breakdown of order (confined to a point or to a line) whose presence induces a strain field that does not vanish even far from the core. For dislocations, what is locally disrupted along the defect line is the inversion symmetry about the atom at the head of the extra half-plane. Other familiar instances of topological defects include vortices in superfluid helium and disclinations in nematic liquid crystals.

Analogously to a point charge in electromagnetism,[1] a topological defect can be revealed by evaluating a suitable field along a closed path or over a closed surface that encloses the defect core. Such defects come equipped with *topological invariants*, namely quantities that characterize the underlying topological structure and remain unchanged under homeomorphisms, i.e., under continuous deformations of the space.[2] These invariants are commonly normalized to integer values—winding numbers, the latter being a canonical example that labels the class of the topological defect.

[1] Note that they also have in common the $1/r$ behaviour of the field.

[2] A typical homeomorphism between two surfaces in $\mathbb{R}^3$ is the one by which a doughnut can be continuously transformed into a coffee mug.

As an illustration, a vortex in a fluid can be formally specified as a mapping from a closed curve $\mathcal{L}$ to the order—parameter space given by the real line segment from 0 to $2m\pi$, with $m = 0, \pm 1, \pm 2, \ldots$ the winding number,

$$\int_{\mathcal{L}} d\theta = \int_{\mathcal{L}} \frac{d\theta}{ds} ds = 2m\pi \tag{6.6}$$

where s denotes the parameter along the closed path $\mathcal{L}$.[3] Equivalently, m counts how many times the loop $\mathcal{L}$ must be traversed for the mapping to return to a single-valued configuration. The integer m is a topological invariant in the sense that any two closed loops with the same m can be smoothly deformed into one another, whereas no continuous deformation can take a map with winding m into one with $m' \neq m$.

Another classic topological invariant is the genus g of a compact, closed surface, which is tied to the surface integral of the Gaussian curvature through the Gauss–Bonnet theorem:

$$\int_S K dA = \int_S (\kappa_1 \kappa_2) dA = 2\pi(2 - 2g) \tag{6.7}$$

where g is the genus of the surface S and $K = \kappa_1 \kappa_2$ is the Gaussian curvature, equal to the product of the two principal curvatures κ_1 and κ_2. By applying Eq. (6.7) to a sphere one immediately finds that the integral evaluates to 4π, hence $g = 0$. For a torus (doughnut), the result is $g = 1$. Consequently, a sphere with $g = 0$ cannot be continuously deformed into a torus or a coffee mug (which have $g = 1$), while the latter two are mutually deformable. Different phases of the same physical system, distinguished by different values of a topological invariant, are separated by a *topological phase transition*.

In solids, topological defects known as dislocations are detected by non-zero values of a loop integral over the microscopic displacement field (i.e. a vector field which tells for each material point its deviation from the ideal crystal lattice position):

$$\oint_{\mathcal{L}} d\mathbf{u} = \oint_{\mathcal{L}} \frac{d\mathbf{u}}{ds} ds = \mathbf{b}, \tag{6.8}$$

where $\mathbf{u}$ plays the same role as θ in Eq. (6.6). A schematic depiction of the physical meaning of the Burgers vector for an edge dislocation in a crystal is shown in Fig. 6.6.

In ordinary crystals, the winding number equals one in lattice-spacing units, so that $|\mathbf{b}| = d$, with d the lattice spacing. In smectic liquid crystals, by contrast,

[3] In the complex plane, the winding number of a closed path $\mathcal{L}$ around the origin is given by $\frac{1}{2\pi i} \oint_{\mathcal{L}} \frac{dz}{z}$.

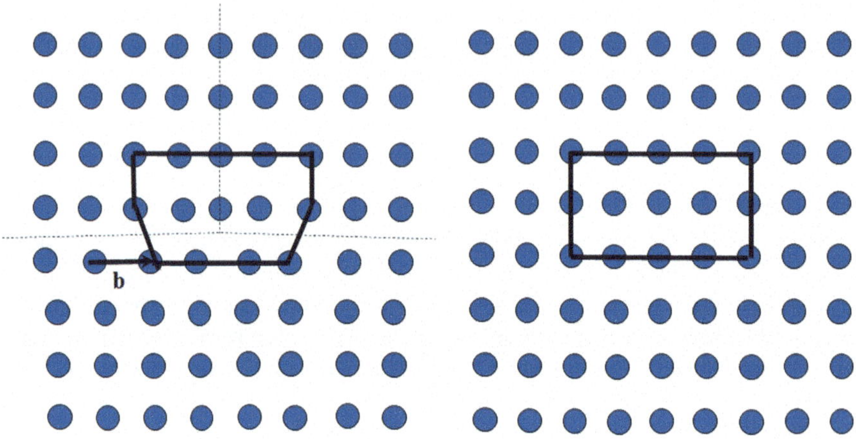

Fig. 6.6 The left panel shows the Burgers vector around an edge dislocation (extra atomic sheet) in a crystal lattice. The Burgers loop around the dislocation is a deformed version of the original rectangular circuit shown in the right panel, when the dislocation is absent. The distortion of the rectangular circuit is measured by the Burgers vector **b**

dislocations may carry winding numbers other than one, and in that case

$$\oint_{\mathcal{L}} d\mathbf{u} = \mathbf{b} = ma\mathbf{e}_z, \tag{6.9}$$

with $m = 0, \pm 1, \pm 2, \ldots$ and $\mathbf{e}_z$ the unit vector along the z-axis.

6.4 Physical Examples

We now illustrate how the abstract notions of topology and differential geometry appear concretely in physical systems, starting from classical elasticity [1].

6.4.1 Compatible and Incompatible Deformations in Solids

The mechanical deformation in a material can be characterized by the displacement vector field, in Cartesian components ($i = x, y, z$) given by u_i [1], which describes the deviations of the material points from the original positions x_i that they have in the undeformed frame to their new positions x_i':

$$x_i' = x_i + u_i\,. \tag{6.10}$$

In the continuum mechanics of solids, if the symmetrized strain tensor of linear elasticity theory [1]: $e_{ij} \equiv \partial_{(i}u_{j)} = \frac{1}{2}(\partial_i u_j + \partial_j u_i)$ obeys the so-called compatibility condition [2, 3]:

$$\nabla \times \nabla \times \mathbf{e} = 0, \tag{6.11}$$

this is equivalent to saying that du_i is a closed differential form, as we shall demonstrate in the following.

The compatibility condition expresses the single-valuedness of the displacement field [4, 5]. For small strain steps, these conditions are equivalent to the fact that the displacements can be obtained by integrating the strains and they can be expressed as

$$\oint_{\mathcal{L}} du_i = 0. \tag{6.12}$$

i.e. the integral of the displacement fields around a close loop $\mathcal{L}$ must vanish. In other words, according to Eq. (6.12), the displacement field u_i is an *exact* differential form. Recall the definitions of exact forms and closed forms that were given in Chap. 3.

Without loss of generality, the displacement field can be written as:

$$du_i = \left(e_{ij} + \omega_{ij}\right) dx_j \tag{6.13}$$

where the first term in bracket is the linearized strain tensor of linear elasticity theory, and is the symmetric part of du_i while the second term is the anti-symmetric part. After some algebraic manipulations, one can prove that:

$$\oint \omega_{ij}\, dx_j = - \int x_l\, \omega_{ij,l}\, dx_j \tag{6.14}$$

and, using standard tensor identities, that:

$$\omega_{ij,l} = \epsilon_{mil}\, \epsilon_{mpq}\, e_{pj,q} \tag{6.15}$$

where $e_{pj,q}$ denotes the derivative along Cartesian component q of the strain tensor components e_{pj}.

We can now write:

$$\oint du_i = \oint \left(e_{ij} - x_l\, \epsilon_{mil}\, \epsilon_{mpq}\, e_{pj,q}\right) dx_j. \tag{6.16}$$

By invoking Stokes' theorem $\oint \mathbf{F} \cdot d\mathbf{x} = \int_S \nabla \times F \cdot \mathbf{n}\, dS$, the above becomes:

$$\oint du_i = -\int \int_S n_r\, \epsilon_{mil} \left(\epsilon_{rsj}\, \epsilon_{mpq}\, e_{pj,qs}\right) x_l\, dS\,. \tag{6.17}$$

The term inside the bracket is then the curl of the curl of the strain tensor:

$$\epsilon_{rsj}\, \epsilon_{mpq}\, e_{pj,qs} \equiv \nabla \times \nabla \times \mathbf{e}\,. \tag{6.18}$$

Therefore, we have just demonstrated that the condition of compatible deformation is equivalent to the requirement

$$\nabla \times \nabla \times \mathbf{e} = 0,$$

which was Eq. (6.11), and that:

$$\nabla \times \nabla \times \mathbf{e} = 0 \;\Leftrightarrow\; \oint_{\mathcal{L}} du_i = 0 \;\Leftrightarrow\; ddu_i = 0. \tag{6.19}$$

This is a direct consequence of the fundamental property of the exterior derivative $dd(\ldots) = 0$, which implies that the exterior derivative of an exact form always vanishes.

This correspondence reveals a deep analogy between elasticity theory and gauge field theories.

In the mechanics of solids, the microscopic displacement field u_i is the 1-form, which plays the same role as the vector gauge potential A in electromagnetism. The fact that $ddu_i = 0$ implies that there are no topological defects in the displacement field of the solid body, just as the identity $ddu_i = 0$ expresses the non-existence of topological defects in the magnetic field, i.e. the non-existence of Dirac magnetic monopoles. Unlike the case of classical electrodynamics, the Bianchi identity for the deformation of a solid can be broken, i.e.

$$ddu_i \neq 0 \tag{6.20}$$

which, thus, also implies:

$$\nabla \times \nabla \times \mathbf{e} \neq 0 \tag{6.21}$$

due to structural disorder or due to the presence of topological defects such as dislocations.

In the presence of disorder, such as in disordered lattices with impurities or in glasses, the total displacement vector u_i can be split into a part called "affine" (A) and a contribution called "nonaffine" (NA) [6]:

$$u_i = u_i^{\mathrm{A}} + u_i^{\mathrm{NA}} = \Lambda_{ki}\, x_k + u_i^{\mathrm{NA}} \tag{6.22}$$

where Λ_{ki} is a matrix of constants naturally related to the macroscopically applied strain. As discussed in Chapter 2 of Ref. [6], see also [7], nonaffine displacements u_i^{NA} arise, in disordered systems, from the need to preserve mechanical equilibrium in the affine positions and throughout the deformation process. In other words, the atoms of the solid first tend to move towards the affine positions since these are the new positions prescribed by the macroscopic strain tensor that describes the deformation. However, due to the disorder breaking the local inversion symmetry around the atom, there is no mechanical equilibrium in the affine position, and there is, instead, a net force acting on the atom which has to be relaxed through an additional "nonaffine" displacement. These nonaffine displacements are chiefly responsible for a negative overall correction to the shear elastic constant in disordered materials, such as glasses [6].

Crucially, while u_i^{A} is a totally "ordered" field, whereby all atoms move in the same direction provided by the macroscopic deformation, the nonaffine displacement field u_i^{NA} is a random field [8]. We can thus already anticipate that, while u_i^{A} contributes nothing to a loop or circulation integral of the displacement field, the nonaffine part u_i^{NA} yields a non-vanishing contribution.

We can write the circulation integral $\oint_{\mathcal{L}} du_i = \ldots$ using the deformation gradient tensor $F_{ij} \equiv \frac{\partial u_i}{\partial x^j}$ as follows:

$$\oint_{\mathcal{L}} \mathbf{F} \cdot d\mathbf{s} = \oint_{\mathcal{L}} \frac{\partial u_i}{\partial x_j} dx^j , \tag{6.23}$$

where the deformation gradient tensor $\mathbf{F}$ has two contributions:

$$F_{ij} = \Lambda_{ij} + \frac{\partial u_i^{\mathrm{NA}}}{\partial x^j} . \tag{6.24}$$

The first term on the right hand side represents the affine contribution, hence it is represented by a matrix of constants, and defines a conservative field $\oint_{\mathcal{L}} \mathbf{\Lambda} \cdot d\mathbf{s} = 0$. This also means that the associated tensor Λ_{ij} is irrotational and the corresponding deformation is a *compatible* deformation.

This fact, in turn, shows that the Burgers vector $\mathbf{b}$ receives no contribution from the affine part of the displacement field, and is, instead, exclusively determined by the nonaffine part of the displacement field u_i^{NA}, and therefore:

$$\oint_{\mathcal{L}} du_i^{\mathrm{NA}} = -b_i . \tag{6.25}$$

This mathematical fact has led to the discovery [9], in molecular simulations, of well-defined topological defects, similar to dislocations, in glasses under deformation. The existence of such well-defined topological defects in glasses, in view of the random atomic structure, had been previously thought to be difficult to identify. These have been subsequently observed also experimentally [10].

6.4.2 Berry Curvature, Chern Number, and Quantum Hall Effect

Topological ideas also play a central role in quantum mechanics, where global properties of parameter spaces give rise to quantized physical observables. A paradigmatic example is provided by Berry curvature and the quantum Hall effect.

Consider a single Dirac cone in two dimensions with a mass gap m, in the dispersion relation of electrons of a 2D material such as e.g. graphene. The Bloch Hamiltonian describing the electrons wavefunctions is

$$H(\mathbf{k}) = d(\mathbf{k}) \cdot \boldsymbol{\sigma}, \qquad d(\mathbf{k}) = \left(vk_x,\ vk_y,\ m\right),$$

where $v > 0$ is a velocity, $\boldsymbol{\sigma} = (\sigma_x, \sigma_y, \sigma_z)$ are Pauli matrices, and $\mathbf{k} = (k_x, k_y) \in \mathbb{R}^2$. The band energies are

$$E_\pm(\mathbf{k}) = \pm\sqrt{v^2k^2 + m^2},$$

and we focus on the lower band $E_-(\mathbf{k})$, assumed to be completely filled so that the system is an insulator.

6.4.2.1 Berry Connection and Berry Curvature

Let $|u_-(\mathbf{k})\rangle$ be a normalized eigenvector of the lower band. The Berry connection (a 1-form in parameter space) is defined by

$$\mathbf{A}(\mathbf{k}) = i\,\langle u_-(\mathbf{k})|\nabla_{\mathbf{k}} u_-(\mathbf{k})\rangle,$$

and the Berry curvature (a 2-form), upon defining

$$\Omega(\mathbf{k}) = \partial_{k_x} A_y - \partial_{k_y} A_x,$$

is

$$\Omega\, dk_x \wedge dk_y.$$

For any two-band Hamiltonian of the form $H = \mathbf{d} \cdot \boldsymbol{\sigma}$, one can use the geometric formula

$$\Omega(\mathbf{k}) = \frac{1}{2}\hat{\mathbf{d}} \cdot \left(\partial_{k_x}\hat{\mathbf{d}} \times \partial_{k_y}\hat{\mathbf{d}}\right), \qquad \hat{\mathbf{d}} = \frac{\mathbf{d}}{|\mathbf{d}|}$$

which expresses the curvature as the pullback of the area form on the Bloch sphere. For the specific $\mathbf{d}(\mathbf{k})$ in this model, a short computation yields

$$\Omega(\mathbf{k}) = \frac{mv^2}{2\,(m^2 + v^2k^2)^{3/2}}\,.$$

6.4.2.2 Chern Number

The first Chern number of the filled band is the total Berry flux through $\mathbf{k}$-space,

$$C = \frac{1}{2\pi}\int_{\mathbb{R}^2} \Omega(\mathbf{k})\, d^2k.$$

Introducing polar coordinates $k = \sqrt{k_x^2 + k_y^2}$ so that $d^2k = k\,dk\,d\phi$, one finds

$$\begin{aligned} C &= \frac{1}{2\pi}\int_0^{2\pi} d\phi \int_0^{\infty} \frac{mv^2}{2\,(m^2 + v^2k^2)^{3/2}}\, k\,dk \\ &= \int_0^{\infty} \frac{mv^2}{2\,(m^2 + v^2k^2)^{3/2}}\, k\,dk. \end{aligned}$$

With the substitution $u = m^2 + v^2k^2$ so that $du = 2v^2k\,dk$, the integral becomes

$$C = \frac{m}{4}\int_{u=m^2}^{\infty} u^{-3/2}\,du = \frac{m}{4}\left[-\frac{2}{\sqrt{u}}\right]_{m^2}^{\infty} = \frac{m}{2}\frac{1}{|m|} = \frac{1}{2}\,\mathrm{sgn}(m).$$

Thus a single massive Dirac cone contributes a half-integer Chern number:

$$C = \tfrac{1}{2}\,\mathrm{sgn}(m).$$

On a lattice, where the Brillouin zone is compact, Dirac cones appear in pairs, and the total Chern number is an integer obtained by summing the contributions of all cones.

6.4.3 *Quantized Hall Conductivity*

For an isolated, completely filled 2D band, the Hall conductivity is proportional to the Chern number,

$$\sigma_{xy} = \frac{e^2}{h}\, C$$

(in units with $\hbar = 1$, equivalently $\sigma_{xy} = (e^2/2\pi) \int \Omega \, d^2k$). For two Dirac cones with the same sign of mass, the total $C = \pm 1$ and the system is a Chern insulator with $\sigma_{xy} = \pm e^2/h$.

6.4.3.1 Differential-Geometric Interpretation

The occupied band defines a complex line bundle over the momentum space (compactified to S^2 in the continuum model or the 2-torus T^2 for a lattice Brillouin zone). The Berry connection $\mathbf{A}$ is a $U(1)$ connection 1-form on this bundle, and the Berry curvature Ω is its curvature 2-form, $\Omega = d\mathbf{A}$. The Chern number

$$C = \frac{1}{2\pi} \int \Omega$$

is the first Chern class of the bundle, a topological invariant. In the Dirac model above, the map $\hat{\mathbf{d}} : \mathbf{k} \mapsto \mathbf{d}/|\mathbf{d}|$ wraps momentum space around the Bloch sphere, and C counts the degree (wrapping number) of this map.

Across classical mechanics, elasticity, electromagnetism, and quantum theory, topology thus provides a unifying language that links global geometric structure to robust physical phenomena.

References

1. L.D. Landau, E.M. Lifshitz, *Theory of Elasticity*, Course of Theoretical Physics, vol. 7, 3rd edn. (Butterworth-Heinemann, Oxford, 1986)
2. A.E.H. Love, *A Treatise on the Mathematical Theory of Elasticity* (Cambridge University Press, Cambridge, 1892)
3. E. Beltrami, Il Nuovo Cimento **20**(1), 5 (1886)
4. A. Acharya, J. Bassani, J. Mech. Phys. Solids **48**(8), 1565 (2000)
5. J.A. Zimmerman, D.J. Bammann, H. Gao, Int. J. Solids Struct. **46**(2), 238 (2009)
6. A. Zaccone, *Theory of Disordered Solids: From Atomistic Dynamics to Mechanical, Vibrational, and Thermal Properties*, 1st edn. Lecture Notes in Physics (Springer, Cham, 2023)
7. A. Zaccone, E. Scossa-Romano, Phys. Rev. B **83**, 184205 (2011). https://doi.org/10.1103/PhysRevB.83.184205. https://link.aps.org/doi/10.1103/PhysRevB.83.184205
8. A. Lemaitre, C. Maloney, J. Stat. Phys. **123**(2), 415 (2006)
9. M. Baggioli, I. Kriuchevskyi, T.W. Sirk, A. Zaccone, Phys. Rev. Lett. **127**, 015501 (2021). https://doi.org/10.1103/PhysRevLett.127.015501. https://link.aps.org/doi/10.1103/PhysRevLett.127.015501
10. V. Vaibhav, A. Bera, A.C.Y. Liu, M. Baggioli, P. Keim, A. Zaccone, Nat. Commun. **16**, 55 (2025)

Chapter 7
Introduction to Group Theory

Abstract This chapter provides an informal introduction to group theory, starting from the canonical four axioms. A strong emphasis is placed on the geometric roots that group theory shares with tensor analysis and differential geometry. These roots lie in the underlying concept of coordinate frame transformations and in the invariance of physical properties under such transformations. The important example of point symmetry groups for Bravais crystal lattices will be introduced and the theorem which fixes the allowed orders of rotation in such lattices will be demonstrated Landau and Lifshitz [1]. Excellent previous textbooks covering these topics are Jones [2] and Cornwell [3]. Group theory provides the natural algebraic language for describing the symmetries of manifolds, fields, and physical systems introduced in the previous chapters.

7.1 Definition of Group

Many of the geometric structures encountered earlier—coordinate changes, rotations, translations, and gauge transformations—form sets closed under composition. Group theory formalizes these sets of transformations and the rules governing their combination.

Definition 7.1 A *group* is a set G equipped with a binary operation (with respect to which G is closed) satisfying the *four axioms*:

1. closure: if $f \in G$ and $g \in G$, then product $fg \in G$;
2. associativity: $f(gh) = (fg)h \; \forall f, g, h \in G$;
3. existence of identity: $\exists \, e \in G$ such that $ge = eg = g \; \forall g \in G$;
4. existence of inverse: $\forall g \in G \, \exists \, g^{-1} \in G$ such that $gg^{-1} = g^{-1}g = e$.

In physics, the elements of a group typically represent transformations acting on space, spacetime, or internal degrees of freedom, while the group operation corresponds to their successive application.

A. Zaccone, *Differential Geometry and Group Theory*,
https://doi.org/10.1007/978-3-032-27518-9_7

7.1.1 Examples

The relevance of groups to geometry and physics becomes clear through a few fundamental examples.

- Translations in $\mathbb{R}^n$ form a group where the binary operation is vector addition of n-dimensional vectors.
- The set of linear transformations in $\mathbb{R}^{3,1}$ that leave invariant the Minkowski quadratic form

$$x_1^2 + x_2^2 + x_3^2 - x_4^2$$

  forms the Lorentz group O(3, 1).
- The set of all transformations in $\mathbb{R}^{3,1}$ that leave invariant the distance between two points in Minkowski spacetime is the *Poincaré group* (which contains the Lorentz group as a subgroup).

Definition 7.2 A group G is called *abelian* if the group operation is commutative.

A key distinction between groups arises from whether the order of transformations matters or not.

In general, the order in which one applies the transformations is important. Indeed, one cannot assume and write that $TT' = T'T$ without a complete knowledge of the group transformations. If $TT' \neq T'T$, then the group is called *non-abelian*.

If, on the other hand, all the elements of a group satisfy the commutative property

$$[T, T'] = TT' - T'T = 0$$

then the group is *abelian*. For example, SO(2) is abelian, because rotations in 2D always commute, while SO(n) with $n \geq 3$ is non-abelian because rotations do not commute in 3D or higher dimensions.

Definition 7.3 Given a group G, the cardinality of the set, denoted sometimes $|G|$ or $[g]$, is called the *order* of the group.

This distinction naturally separates discrete symmetries from continuous ones, a difference that plays a crucial role in both crystallography and field theory.

7.1.2 Examples: Z_n and S_n

Finite groups often arise from discrete symmetries such as reflections, permutations, and rotations by fixed angles.

- We define Z_n as the set of integers modulo n, which forms a finite Abelian group of order n. The parity group $\mathbb{Z}_2$ contains two elements, 1 and -1, which are applied under ordinary multiplication.

  For example, the parity transformation of a function $f(x)$ is defined as:

$$f(x) \Rightarrow f(-x)$$

  or, in the case of the reflection of the function:

$$f(x) \Rightarrow -f(x)$$

  $\mathbb{Z}_2$ is finite, has order 2, and is abelian.

  For any positive integer m, one defines the phase

$$\exp\left(i\frac{2\pi k}{m}\right)$$

  which gives the k-th element of the cyclic group $\mathbb{Z}_m$, abelian of order m.
- We define S_n as the group of permutations of $\{1, \ldots, n\}$ objects.

Definition 7.4 A *Lie group* is a group in which each element depends continuously on one or more parameters:

$$g = g(\{\alpha\})$$

Each element g depends continuously on a parameter α or on a set of parameters α_k. These groups are of infinite order.

Lie groups thus combine algebraic structure with the smooth manifold structure introduced in the previous chapter.

For example, $\mathrm{O}(n)$, $\mathrm{SO}(n)$, $\mathrm{U}(n)$ and $\mathrm{SU}(n)$ are all Lie groups of infinite order.

Definition 7.5 A matrix group $G \subseteq \mathrm{GL}(n, \mathbb{C})$ is called *compact* if it is a compact subset of $\mathbb{C}^{n\times n}$ with respect to the standard topology (equivalently, if it is closed and bounded).

A convenient norm on matrix space is the Hilbert–Schmidt norm

$$\|A\|^2 = \mathrm{Tr}(A^\dagger A).$$

Bounding $\mathrm{Tr}(A^\dagger A)$ therefore guarantees boundedness of the set

7.1.3 *Groups of Matrices*

In practice, many symmetry transformations are represented by matrices acting linearly on vectors or fields.

Matrices form groups (under matrix multiplication) in a natural way, since every set $\{D\}$ of non-singular $n \times n$ matrices that contains the inverse D^{-1} for each element and the identity I automatically satisfies the properties of axioms 2–4.

Only axiom 1 (closure under the group multiplication law) remains uncertain. A set of non-singular matrices forms a group if the product of any pair remains in the set. This usually happens if all the matrices leave some property invariant (*linear transformations*). This is, typically, the case of physical transformations that are represented via matrices.

Definition 7.6 The *general linear group* over a field $\mathbb{K}$ is the group $\mathrm{GL}(n, \mathbb{K})$ consisting of all non-singular matrices in $\mathbb{K}^{n\times n}$.

Subgroups (see the next chapter for the formal definition of subgroup) of $\mathrm{GL}(n, \mathbb{K})$ are obtained by imposing additional geometric constraints, such as the preservation of lengths or angles.

Proposition 7.1 $\mathrm{GL}(n, \mathbb{K})$ *is an infinite, non-Abelian, non-compact group.*

Definition 7.7 The *special linear group* is defined as

$$\mathrm{SL}(n, \mathbb{K}) \equiv \{\mathbf{A} \in \mathrm{GL}(n, \mathbb{K}) : \det \mathbf{A} = 1\}.$$

7.2 Orthogonality and Rotations

Rotational symmetry plays a central role in geometry and physics, from rigid body motion to spacetime symmetries.

We have said that orthogonal groups leave quadratic forms invariant. This is verified when

$$A^{\mathrm{T}} A = I$$

which is the definition of an orthogonal matrix.

$$A^{\mathrm{T}} = A^{-1}$$

From this it follows that

$$|A^{\mathrm{T}} A| = |A^{\mathrm{T}}||A| = |A|^2 = |I| = 1$$

where we used that $|A^{\mathrm{T}}| = |A|$. Therefore, $\det(A) = \pm 1$.

Rotations in n-dimensional space are described by the group SO(n), which contains orthogonal matrices with determinant $= +1$ in n dimensions.

Matrices that instead belong to O(n) but have det $= -1$ represent improper rotations (while the "proper" rotations described by SO(n) are the only rotations that can be applied in a rigid body). Only proper rotations correspond to physically realizable rigid motions without reflection.

7.2.1 Example: Proper and Improper Rotations

The prototype of improper rotations is given by reflections such as:

$$\mathbf{r}' = -\mathbf{r}, \qquad \mathbf{r}' = R(I)\mathbf{r}$$

where the rotation matrix is

$$R(I) = \begin{pmatrix} -1 & 0 & 0 \\ 0 & -1 & 0 \\ 0 & 0 & -1 \end{pmatrix}$$

where I, here, denotes the inversion. Clearly, we have that:

$$\det R(I) = -1, \qquad R(I) \in \mathrm{O}(3)$$

Instead, a proper rotation of an angle θ about a cartesian axis is given by:

$$\mathbf{r}' = R(\theta)\mathbf{r}, \qquad \mathbf{r}' = \begin{pmatrix} x' \\ y' \\ z' \end{pmatrix}, \qquad \mathbf{r} = \begin{pmatrix} x \\ y \\ z \end{pmatrix}$$

Rotation by an angle θ around, e.g., the x-axis corresponds to the following transformation:

$$\begin{cases} x' = x \\ y' = y\cos\theta + z\sin\theta \\ z' = -y\sin\theta + z\cos\theta \end{cases}$$

which identifies the rotation matrix $R(\theta)$:

$$R(\theta) = \begin{pmatrix} 1 & 0 & 0 \\ 0 & \cos\theta & \sin\theta \\ 0 & -\sin\theta & \cos\theta \end{pmatrix}$$

which clearly satisfies:

$$\det R(\theta) = 1, \qquad R(\theta) \in \mathrm{SO}(3).$$

This, of course, holds for rotations around any axis by any value of the rotation angle θ. Furthermore, any matrix obtained as a product $R(I)R(\theta)$ is an improper rotation with determinant -1 and $R(I)R(\theta) \in \mathrm{O}(3)$.

Definition 7.8 The *orthogonal group* is the group $\mathrm{O}(n)$ of orthogonal matrices in $\mathbb{R}^{n\times n}$ with $\det(R) = \pm 1$.

Definition 7.9 The *special orthogonal group* is

$$\mathrm{SO}(n) \equiv \{\mathbf{A} \in \mathrm{O}(n) : \det \mathbf{A} = 1\}.$$

$\mathrm{O}(n)$ is the group of both proper and improper rotations in $\mathbb{R}^n$, whereas $\mathrm{SO}(n)$ is the subgroup of proper rotations.

$\mathrm{O}(n)$ is the group of matrices that leave invariant the real quadratic form $x_1^2 + \cdots + x_n^2$:

$$(\mathbf{y}, \mathbf{y}) = \mathbf{y}^\mathsf{T}\mathbf{y} = \mathbf{x}^\mathsf{T}\mathbf{A}^\mathsf{T}\mathbf{A}\mathbf{x} = \mathbf{x}^\mathsf{T}\mathbf{I}_n\mathbf{x} = \mathbf{x}^\mathsf{T}\mathbf{x} = (\mathbf{x}, \mathbf{x}).$$

Proposition 7.2 *If* $\mathbf{A} \in \mathrm{O}(n)$, *then* $\det \mathbf{A} = \pm 1$.

Proof From $\mathbf{A}^\mathsf{T}\mathbf{A} = \mathbf{I}_n$ it follows that $1 = \det(\mathbf{A}^\mathsf{T}\mathbf{A}) = \det(\mathbf{A})^2$. □

Definition 7.10 The *unitary group* is the group $\mathrm{U}(n)$ of unitary matrices in $\mathbb{C}^{n\times n}$.

Definition 7.11 The *special unitary group* is

$$\mathrm{SU}(n) \equiv \{\mathbf{A} \in \mathrm{U}(n) : \det \mathbf{A} = 1\}.$$

All complex $n \times n$ matrices that leave invariant the quadratic form

$$x_1^* x_1 + x_2^* x_2 + \cdots + x_n^* x_n$$

constitute the group $\mathrm{U}(n)$. Those with determinant equal to $+1$ form the special unitary group $\mathrm{SU}(n)$.

7.2.2 *Examples: Unitary Groups*

In quantum mechanics, symmetry transformations act on complex-valued vector spaces and are naturally described by unitary groups.

- $\mathrm{U}(1)$ describes rotations in the Argand plane.

- SU(3) is the symmetry group of the strong interaction (in Quantum Chromo Dynamics).
- SU(5) and SU(10) are used in high-dimensional grand unification theories in an attempt to unify the electroweak and the strong force.

7.3 Crystallography

Discrete symmetry groups play a fundamental role in solid-state physics, where the atomic structure of matter imposes strong geometric constraints.

Definition 7.12 A *crystal* or *Bravais lattice* is defined as a lattice of atoms invariant under translations, cf. Fig. 7.1.

Definition 7.13 The *space group* of a crystal is the group of transformations (translations, rotations, reflections and inversions) that leave it invariant.

Definition 7.14 The *point group* or *symmetry group* of a crystal is the subgroup of its space group obtained by excluding translations.

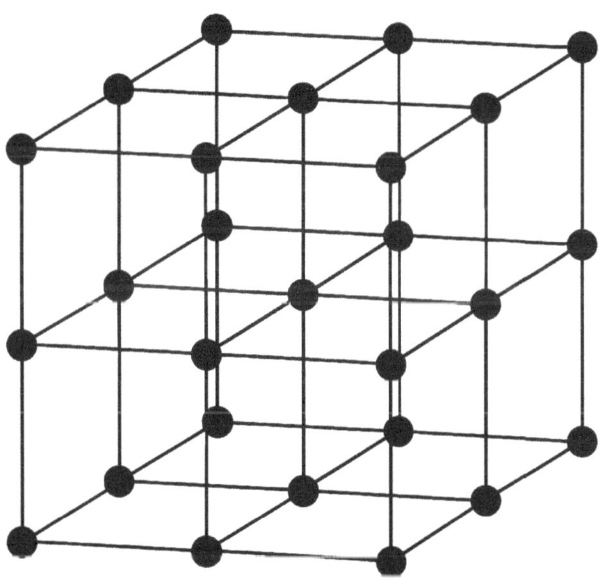

Fig. 7.1 Schematic of a 3D simple cubic Bravais lattice obtained by an ordered repetition of a primitive cell which is just a cube with atoms occupying its vertices. For simplicity only a few cells are shown

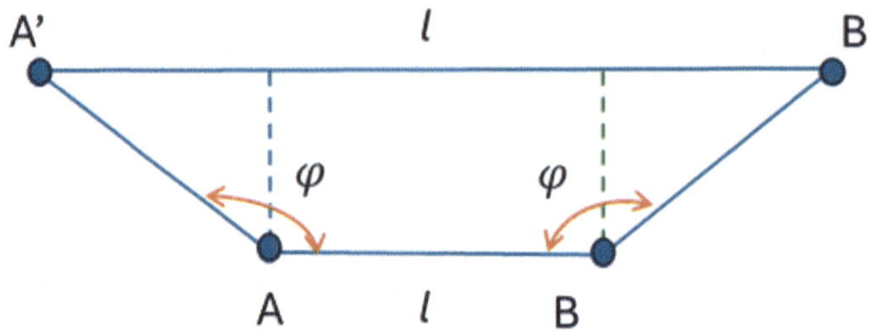

Fig. 7.2 Schematic of a section of a Bravais lattice, with A and B being lattice points (atoms on the lattice), needed to demonstrate the crystallographic restriction theorem 7.1

7.3.1 Physical Example: The Bravais Lattices

The simple cubic Bravais lattice has the largest symmetry group: it is invariant under all permutations of the coordinates with all possible sign choices, hence its point group has order $2^3 3! = 48$.

Group-theoretical arguments place strong restrictions on the rotational symmetries compatible with the translational order of the crystal lattice.

Theorem 7.1 (Crystallographic Restriction Theorem) *The only possible rotation angles for Bravais lattices are $\varphi_n = \frac{2\pi}{n}$ with $n = 1, 2, 3, 4, 6$.*

Proof With reference to Fig. 7.2, let A be a lattice point (an atom) lying on an axis of symmetry perpendicular to the plane considered, and let B be a point at distance l from A obtained via a translation of the lattice; since the lattice is periodic along the direction from A to B, atom B also lies on an axis of symmetry perpendicular to the plane.

Let A' and B' be the images of A and B obtained by a rotation through an angle $\varphi_n = \frac{2\pi}{n}$ about B and A, respectively. Assuming that the rotation by φ_n is a symmetry of the lattice, the points A' and B' must belong to the lattice and can be made to coincide by a translation.

If l is the smallest translation period of the lattice, we must have $A'B' = pl$ with $p \in \mathbb{N}$, that is

$$A'B' = l + 2l \sin\left(\varphi_n - \frac{\pi}{2}\right) = l\,(1 - 2\cos\varphi_n) = pl \quad \rightarrow \quad \cos\varphi_n = \frac{1}{2}(1 - p).$$

From the condition $|\cos\varphi_n| \leq 1$ it follows that the possible values of p are $p = 0, 1, 2, 3$, corresponding to $n = 1, 2, 3, 4, 6$. □

As a consequence, one can have, for instance, a hexagonal lattice, but not a pentagonal one. In fact, "quasi-crystals" with $n = 5$ were discovered in 1982 by Israeli physicist Dan Schechtman: these structures exhibit long-range orientational order but not long-range translational symmetry.

7.4 The Cyclic Group C_n

The rotational symmetries of crystals and molecules are naturally described by cyclic and dihedral groups.

C_n is the group of rotations by an angle $\frac{2\pi}{n}$ around an axis of symmetry. C_n is a cyclic group. C_n^2 denotes a rotation by $\frac{4\pi}{n}$, which can be obtained by repeating the rotation C_n^1 twice.

In general, C_n^m is obtained by repeating the rotation C_n^1 m times, with

$$C_n^m = (C_n^1)^m, \qquad \text{where } C_n^1 \equiv C.$$

Here, C_n^1 is a rotation by $\frac{2\pi}{n}$.

It is clear that

$$C_n^n = E.$$

C_n is a finite cyclic group generated by the element C:

$$C_n = \{E, C, C^2, \ldots, C^{n-1}\}.$$

It is abelian:

$$C^r C^s = C^{r+s} = C^{s+r}.$$

The correspondence

$$C^r \longleftrightarrow r \pmod{n}$$

defines an isomorphism between C_n and the additive group $\mathbb{Z}_n$.

Proposition 7.3 $C_n \cong Z_n$.

That is, the multiplication of elements of C_n is equivalent to adding the exponents of C, but modulo n, since an exponent n is identical to an exponent 0.

$$C^r C^s = C^{\,r+s \pmod{n}}.$$

Here Z_n is understood under the composition law given by addition.

Proposition 7.4 *The point group D_n has order $2n$ and C^n is a subgroup of it.*

Including reflections in addition to rotations leads to a richer class of symmetry groups.

7.4.1 Examples: Dihedral Groups

The point group D^4 of a regular polygon with 4 sides (a square) is composed of: the rotations about the axis in O orthogonal to the plane, and the dihedral rotations about in-plane symmetry axes, cf. Fig. 7.3. Let c be the rotation by an angle $\pi/2$ (counterclockwise). Then

$$c^4 = E$$

that is, a rotation of $\pi/2$ about the axis orthogonal to the plane repeated 4 times returns the identity E. Thus, this is a rotation of order 4.

Furthermore, there are rotations of order 2, that is rotations of π, around the axes of symmetry in the plane (cf. Fig. 7.3):

$$b_1 \text{ around the axis } OX, \qquad b_2 \text{ around the axis } AC,$$

$$b_3 \text{ around the axis } OY, \qquad b_4 \text{ around the axis } BD.$$

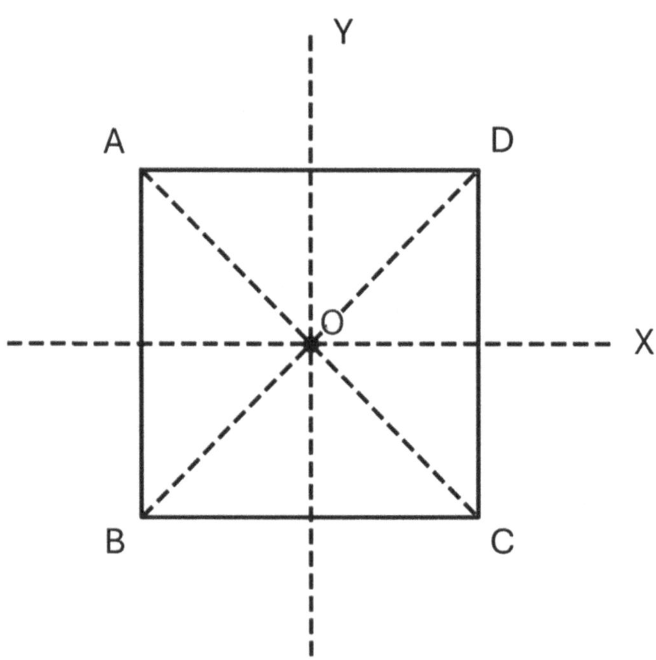

Fig. 7.3 Schematic needed to define the symmetry operations of the group D_4

Verify that:

$$b_2 = bc, \qquad b_3 = bc^2, \qquad b_4 = bc^3,$$

where $b = b_1$.

Therefore, the group is generated by the elements b and c.

$$D_4 = \text{gp}\{c, b\}, \qquad c^4 = b^2 = (bc)^2 = E.$$

Here, gp stands for "generated by". Generated means that every element of the group can be obtained as a product of the elements b and c (raised to certain powers).

All this can be easily generalized to order n: D_n is the symmetry group of a regular polygon with n sides. It contains rotations of order n around the axis passing through the center and orthogonal to the plane and it also contains n rotations of order 2 around the symmetry axes in the plane.

Thus,

$$D_n = \text{gp}\{c, b\}, \qquad c^n = b^2 = (bc)^2 = E.$$

The group has order:

$$|D_n| = 2n.$$

For example, D_4 has 8 elements:

$$D_4 = \{\, E,\ c,\ c^2,\ c^3,\ b,\ bc,\ bc^2,\ bc^3 \,\}.$$

In general:

$$D_n = \{\, E,\ c,\ c^2,\ \ldots,\ c^{n-1},\ b,\ bc,\ \ldots,\ bc^{n-1} \,\}.$$

Note The rotations c^r (with $r = 1, \ldots, n-1$) with $c^r \in C_n$ form a subgroup of D_n. The subset of elements

$$\{\, c,\ c^2,\ \ldots,\ c^{n-1} \,\}$$

is closed under the group operations.

Every product of rotations around the axis orthogonal to the plane gives another rotation around that same axis.

7.4.2 Example: Multiplication Table of C_3

The multiplication table of group C_3 can be constructed as follows:

	e	c	c^2
e	e	c	c^2
c	c	c^2	e
c^2	c^2	e	c

Group theory thus provides the unifying framework for describing symmetry across geometry, topology, and physical systems, from spacetime transformations to crystal lattices and quantum phases.

References

1. L.D. Landau, E.M. Lifshitz, *Statistical Physics, Part 1*, Course of Theoretical Physics, vol. 5, 3rd edn. (Pergamon Press, Oxford, 1980)
2. H.F. Jones, *Groups, Representations and Physics*, 2nd edn. (Taylor & Francis, Bristol, 1998)
3. J.F. Cornwell, *Group Theory in Physics*, vol. 1 (Academic Press, London, 1997)

Chapter 8
The Algebra of Groups

Abstract This chapter introduces the basic concepts, definitions and tools of the algebra of groups. The central result is the Lagrange theorem which opens up the way for concepts such as the quotient group, of fundamental importance in modern physics. Standard references for this part are the books by Jones [1] and by Cornwell [2]. While the previous chapter focused on concrete symmetry groups and their geometric realizations, the present chapter develops the abstract algebraic structure that underlies all group-based symmetry arguments in physics.

8.1 Subgroups

Once a symmetry group has been identified, it is often necessary to analyze its internal structure by identifying smaller groups contained within it.

Definition 8.1 Given a group G, a subset H of G is called a *subgroup* if it is a group under the same operation as G. One writes $H \leq G$.

Physically, subgroups correspond to restricted sets of symmetry operations that remain valid when additional constraints are imposed.

Definition 8.2 A subgroup H of G is called a *proper subgroup* if $H \neq G$. One writes $H < G$.

Example
D_3 has proper subgroups $B_2 = \{e, b\}$ and $C_3 = \{e, c, c^2\}$.

To understand how a subgroup sits inside a larger group, it is useful to partition the group into equivalent subsets.

A. Zaccone, *Differential Geometry and Group Theory*,
https://doi.org/10.1007/978-3-032-27518-9_8

Definition 8.3 Given a subgroup H of a group G, the *left coset* of an element $g \in G$ is defined as

$$gH \equiv \{gh \in G : h \in H\}.$$

Definition 8.4 Given a subgroup H of a group G, the *right coset* of an element $g \in G$ is defined as

$$Hg \equiv \{ hg \in G : h \in H \}.$$

In general, left and right cosets need not coincide. However, if H is a normal subgroup of G, then

$$gH = Hg \qquad \forall g \in G.$$

This property ensures that the multiplication of cosets is well defined.

Definition 8.5 An *equivalence class* of a group G is defined as a set of mutually conjugate elements of G, namely

$$g_1 \sim g_2 \;\Leftrightarrow\; \exists f \in G : g_2 = f g_1 f^{-1}.$$

Conjugacy captures the idea that some group elements represent the same symmetry operation viewed from different reference frames.

Example
In D_4, the elements b and bc^2 are mutually conjugate since they correspond to dihedral rotations by the same angle π, whose axes can be made to coincide by a rotation. Physically, this corresponds to the fact that these rotations can be made equivalent by a rigid-frame rotation.

8.2 Lagrange's Theorem and the Quotient Group

A central question in group theory is how the size of a group is related to the size of its subgroups.

Theorem 8.1 (Lagrange) *Given a finite group G and a subgroup $H \leq G$, the order $|H|$ divides the order $|G|$.*

Proof The proof is a classic example of *reductio ad absurdum*. Given $g \in G$ such that $g \notin H$, one has $gH \cap H = \emptyset$. Indeed, by contradiction, suppose there exist $g \in G \setminus H$ and $h_1, h_2 \in H$ such that $gh_1 = h_2$. Then

$$gh_1 = h_2 \Rightarrow (gh_1)h_1^{-1} = h_2 h_1^{-1} \Rightarrow g = h_2 h_1^{-1} \in H,$$

By inserting the identity in the form $g_b g_b^{-1}$, this can be rewritten as

$$(g_a h_i)(g_b h_j) = g_a g_b \left(g_b^{-1} h_i g_b \right) h_j.$$

Since H is a normal subgroup, one has

$$g_b^{-1} h_i g_b \in H,$$

and since H is closed under the group operation, the product

$$h_k \equiv (g_b^{-1} h_i g_b) h_j$$

is again an element of H. Therefore,

$$(g_a h_i)(g_b h_j) = g_c h_k, \qquad \text{with } g_c = g_a g_b.$$

The requirement that H be a normal subgroup is essential. Indeed, if H were not normal, the product of two cosets could depend on the choice of representatives g_a and g_b. Normality ensures that

$$g_b^{-1} h_i g_b \in H,$$

so that the resulting element always belongs to the coset $(g_a g_b)H$ independently of the representatives chosen.

This shows that the product of two elements belonging to the cosets $g_a H$ and $g_b H$ is an element of the coset $(g_a g_b)H$. In particular, the multiplication law between cosets is well defined and given by

$$(g_a H)(g_b H) = (g_a g_b)H.$$

This multiplication law mirrors the original group operation while effectively "factoring out" the internal structure of the subgroup.

The set of left cosets of an invariant (normal) subgroup H is closed under this multiplication and therefore forms a group. If G is a group and $H \trianglelefteq G$, one can thus construct a new group whose elements are the left cosets gH; this group is called the *quotient group* and is denoted by G/H.

Proposition 8.1 *The set G/H is a group under the operation*

$$(g_1 H)(g_2 H) \equiv (g_1 g_2)H.$$

Proposition 8.2 *G/H is a group with multiplication law*

$$(g_1 h_1)(g_2 h_2) = (g_1 g_2) h_3.$$

Proof

$$(g_1h_1)(g_2h_2) = g_1(g_2g_2^{-1})h_1g_2h_2 = g_1g_2(g_2^{-1}h_1g_2)h_2 = g_1g_2h_1'h_2 = (g_1g_2)h_3.$$

□

Proposition 8.3

$$|G/H| = \frac{|G|}{|H|}.$$

Proof By Lagrange's theorem. □

This proposition shows that the number of elements of the quotient group G/H is equal to the number of distinct left cosets of H in G, also called the *index* of H in G. The quotient group therefore measures how many distinct symmetry classes remain once the subgroup symmetries are identified as equivalent.

By Lagrange's theorem, all left cosets of H have the same cardinality as H and form a partition of the group G. Since G is the disjoint union of these cosets, the total number of elements of G is the product of the number of cosets and the number of elements in each coset. Therefore, the order of the quotient group is given by

$$|G/H| = \frac{|G|}{|H|}.$$

This result holds for any finite group G and any subgroup $H \leq G$. The normality of H is not required for the counting argument, but it is necessary in order for G/H to carry a well-defined group structure.

8.4 Direct Product Group

In some cases, the structure of a group can be completely decomposed into independent symmetry sectors.

Definition 8.8 Let $A, B \trianglelefteq G$. The group G is the *direct product* of A and B, written

$$G = A \times B,$$

if

$$A \cap B = \{e\}, \qquad AB = G,$$

and every element of A commutes with every element of B.

This definition formalizes the idea that a group G can be decomposed into two independent subgroups A and B. Direct products thus formalize the notion that different symmetries may act independently on a physical system.

The condition $ab = ba$ for all $a \in A$ and $b \in B$ ensures that elements of the two subgroups commute, so that the order in which they are multiplied is irrelevant. This allows elements of A and B to be combined without ambiguity.

The second condition, requiring that every element $g \in G$ can be written as $g = ab$ with $a \in A$ and $b \in B$, guarantees that the subgroups together generate the whole group. In this case, the group G is completely characterized by the independent actions of A and B.

When both conditions are satisfied and the decomposition is unique, the group G behaves as the product of two separate symmetry groups, and is therefore called the direct product of A and B.

Example

A fundamental example of a direct product decomposition is provided by the orthogonal group $\mathrm{O}(n)$, and the decomposition thereof into a direct product of proper rotations times space inversions.

Every orthogonal matrix $A \in \mathrm{O}(n)$ satisfies $\det A = \pm 1$. The subgroup

$$\mathrm{SO}(n) = \{A \in \mathrm{O}(n) : \det A = +1\}$$

consists of all proper rotations, while the discrete subgroup

$$\{\mathbf{I}_n, -\mathbf{I}_n\} \cong \mathbb{Z}_2$$

accounts for the sign of the determinant and represents the presence or absence of a reflection.

The matrix $-\mathbf{I}_n$ commutes with every element of $\mathrm{SO}(n)$, and every orthogonal matrix can be written uniquely either as R or as $(-\mathbf{I}_n)R$, with $R \in \mathrm{SO}(n)$. Therefore, every element of $\mathrm{O}(n)$ can be expressed as the product of an element of $\mathrm{SO}(n)$ and an element of $\{\mathbf{I}_n, -\mathbf{I}_n\}$.

Since the two subgroups commute and together generate the entire group, $\mathrm{O}(n)$ satisfies the conditions of Definition 8.7 and can be written as the direct product

$$\mathrm{O}(n) = \mathrm{SO}(n) \otimes \{\mathbf{I}_n, -\mathbf{I}_n\}.$$

This decomposition shows that the continuous rotational degrees of freedom described by $\mathrm{SO}(n)$ are independent of the discrete reflection symmetry encoded in $\{\mathbf{I}_n, -\mathbf{I}_n\}$.

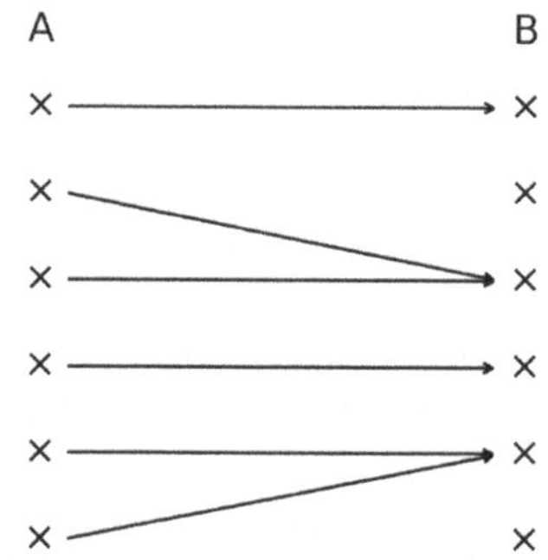

Fig. 8.2 Schematic representation of a homomorphism $f : A \to B$. Each element of A is mapped to a unique element of B, but different elements of A may have the same image in B, and some elements of B may have no preimage. The set of elements of B that are images of elements of A constitutes the image $\text{Im}\, f \subseteq B$

8.5 Mappings, Homomorphisms and Isomorphisms

To compare different groups and relate abstract symmetry structures to concrete realizations, one needs structure-preserving mappings.

A homomorphism is a mapping f from a set A to a set B

$$f : A \to B$$

that preserves a certain structure. Every element of A is mapped to a unique image belonging to B

$$\big(f(a) \in B\big)$$

but the inverse process is not necessarily true. The homomorphism is **not** one-to-one.

In physics, homomorphisms often relate an abstract symmetry group to its action on physical states or observables.

Only the elements of B that are images of elements of A constitute the image.

$$\text{Im}\, f \quad \text{or} \quad f(A)$$

$$\text{Im}\, f = \{b \in B \mid b = f(a),\ a \in A\}$$

Thus, every element of A is mapped to a unique image $b = f(a)$, but an element $b \in B$ can:

(i) be the image of several elements of A,
(ii) not be the image of any element of A.

A schematic conceptual diagram of a homomorphism is shown in Fig. 8.2.

A mapping is **injective** if it is one-to-one (1:1), but there exist elements of the codomain that do not have a corresponding element in the domain. It is **surjective** if all elements of the codomain have a corresponding element in the domain, but it is not one-to-one. It is **bijective** if it is both injective and surjective.

A **homomorphism** is a mapping between groups that preserves the group operation. An **isomorphism** is bijective.

8.5.1 Homomorphism

A homomorphism is a mapping between groups

$$f : A \to B$$

such that for all $a_1, a_2 \in A$,

$$f(a_1 a_2) = f(a_1) f(a_2).$$

In words: the image of the product must be equal to the product of the images.

NB $a_1 a_2$ follows the multiplication law of A, whereas $f(a_1) f(a_2)$ follows the multiplication law of B.

In the case of a homomorphism, Im $f(A)$ is a subgroup of B.

8.5.2 Kernel

The kernel of a homomorphism identifies elements of the original group that act trivially in the target space.

$$\ker f = \{ a \in A \mid f(a) = E_B \in B \}$$

It can be shown that the kernel is also a subgroup of A; in particular, it is a **normal subgroup**.This observation foreshadows the deep connection between homomorphisms and quotient groups.

Indeed,

$$f(gkg^{-1}) = f(g) f(k) f(g^{-1}) = f(g) E_B (f(g))^{-1} = E_B,$$

where to arrive at the final result we used that fact that:

$$f(a) = f(a E_A) = f(a) f(E_A)$$

which implies

$$f^{-1}(a)f(a) = f(E_A) = E_B.$$

Example
Verify the following mapping and determine whether it is a homomorphism.

$$f : D_4 \to \mathbb{Z}_2$$

The mapping is defined by rotation elements

$$f(e) = f(c^2) = f(b) = f(bc^2) = +1$$

$$f(c) = f(c^3) = f(bc) = f(bc^3) = -1.$$

Recall that D_4 is generated by b and c, and therefore it is sufficient to check the homomorphism property on these generators.

We verify the defining condition of a homomorphism,

$$f(gh) = f(g)f(h),$$

for representative elements, by induction.

For example,

$$f(bc^3) = f(b)f(c^3).$$

Using the definition of f,

$$f(b) = +1, \qquad f(c^3) = -1,$$

and hence

$$f(b)f(c^3) = (+1)(-1) = -1,$$

which agrees with

$$f(bc^3) = -1.$$

Therefore, the mapping preserves the group operation and defines a homomorphism from D_4 to $\mathbb{Z}_2$.

8.6 Representation of a Group G

The most important applications of group homomorphisms in physics arise when groups are represented as transformations acting on vector spaces.

A particularly important homomorphism in physics is one that maps a group G into a group of non-singular $n \times n$ matrices, with matrix multiplication as the law of composition. In this case, the matrix group provides an n-dimensional representation of G.

Importantly If the homomorphism is an **isomorphism** (bijective, i.e. one-to-one and onto), then the representation is said to be **faithful**.

Example
The group O(3) is isomorphic to the direct product

$$\mathrm{O}(3) \cong \mathrm{SO}(3) \rtimes \mathbb{Z}_2.$$

where $\mathbb{Z}_2$ is the group (of order 2) consisting of the matrices

$$\mathbf{I}_3 \quad \text{and} \quad -\mathbf{I}_3.$$

Thus, in general $\mathrm{O}(n)$ is a semidirect product of $\mathrm{SO}(n)$ with $\mathbb{Z}_2$. Indeed, O(3) is isomorphic to the group of all rotations in three dimensions, while SO(3) is isomorphic to the subgroup of proper rotations. This implies that the group of all rotations is isomorphic to the direct product of the group of proper rotations and the group

$$\{E, I\},$$

consisting of the identity and the spatial inversion operator.

The algebraic structures developed in this chapter provide the foundation for understanding symmetry breaking, gauge equivalence, and representations in both classical and quantum physics.

References

1. H.F. Jones, *Groups, Representations and Physics*, 2nd edn. (Taylor & Francis, Bristol, 1998)
2. J.F. Cornwell, *Group Theory in Physics*, vol. 1 (Academic Press, London, 1997)

Chapter 9
The Lie Theory of Rotations

Abstract After having laid down the algebraic foundations of group theory, we now turn to an informal description of rotations introducing Lie's idea on how to algebraically describe rotations. This idea is of great importance in physics, where it provides a connection between infinitesimal rotations (the Lie generators) and finite rotations. The power of this approach becomes evident when it is applied to higher dimensional geometric spaces where traditional tools such as our direct intuition and trigonometry cannot be used or trusted. An excellent reference for this part is the book by Zee [1]. This chapter provides the infinitesimal counterpart to the finite symmetry groups discussed previously, revealing how continuous symmetries are encoded in their generators and algebraic structure.

9.1 Lie Groups

Having developed the algebraic structure of groups and their representations, we now specialize to groups that depend continuously on one or more parameters.

9.1.1 2D Rotations

Rotations provide the simplest and most intuitive example of a continuous symmetry group.

It is well known that in $\mathbb{R}^2$ a rotation of the Cartesian reference frame by an angle $\theta \in [0, 2\pi)$ is given by $\mathbf{r}' = \mathbf{R}(\theta)\mathbf{r}$, with

$$\mathbf{R}(\theta) = \begin{bmatrix} \cos\theta & \sin\theta \\ -\sin\theta & \cos\theta \end{bmatrix} \tag{9.1}$$

A. Zaccone, *Differential Geometry and Group Theory*,
https://doi.org/10.1007/978-3-032-27518-9_9

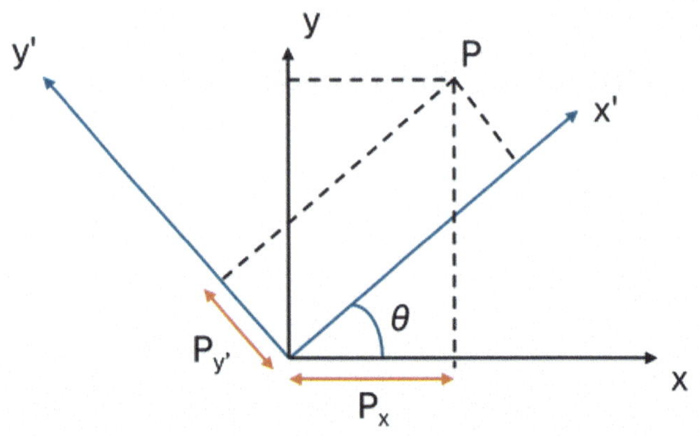

Fig. 9.1 Geometric construction used to derive the two-dimensional rotation matrix Eq. (9.1) from elementary trigonometry. A vector in the plane is rotated by an angle θ with respect to the original Cartesian axes; its projections onto the rotated axes give the trigonometric relations that lead to the matrix representation of a planar rotation

It is immediately seen that $|\mathbf{r}'|^2 = |\mathbf{r}|^2$; rotations can in fact be characterized by this property, which corresponds to imposing

$$\mathbf{R}^{\mathsf{T}}(\theta)\mathbf{R}(\theta) = \mathbf{I}_2 \tag{9.2}$$

that is, $\mathbf{R}(\theta) \in \mathrm{O}(2)$. It is easy to see that the group operation is the composition of rotations $\mathbf{R}(\theta_1)\mathbf{R}(\theta_2) = \mathbf{R}(\theta_1 + \theta_2)$, with identity $\mathbf{R}(0)$ and inverse $\mathbf{R}(-\theta)$.

It is important to recall that $\mathrm{O}(n)$ is the group of both proper and improper rotations: the proper rotations described in Eq. (9.1) satisfy the additional condition $\det \mathbf{R} = 1$ and form the subgroup $\mathrm{SO}(n)$.

Equation (9.1) can be derived either from trigonometric relations (cf. the derivation in [2] following the scheme in Fig. 9.1) or by reasoning directly from the condition in Eq. (9.2).

Rather than relying on geometric intuition or trigonometric constructions, Lie's approach focuses on the behavior of transformations arbitrarily close to the identity.

Following the latter approach, which coincides with Lie's point of view, the basic idea is to consider infinitesimal rotations, i.e. rotations with $\theta \ll 1$:

$$\mathbf{R}(\theta) \simeq \mathbf{I}_2 + \theta\mathbf{A} + o(\theta^2) \tag{9.3}$$

The orthogonality condition then becomes

$$\mathbf{R}^{\mathsf{T}}(\theta)\mathbf{R}(\theta) \simeq \left(\mathbf{I}_2 + \theta\mathbf{A}^{\mathsf{T}}\right)(\mathbf{I}_2 + \theta\mathbf{A}) \simeq \mathbf{I}_2 + \theta(\mathbf{A}^{\mathsf{T}} + \mathbf{A}) = \mathbf{I}_2$$

that is,

$$\mathbf{A}^{\mathsf{T}} = -\mathbf{A} \tag{9.4}$$

Matrices satisfying this condition generate infinitesimal rotations and encode the local structure of the rotation group.

In $\mathrm{GL}(2, \mathbb{R})$ the space of antisymmetric matrices is one-dimensional. Therefore every matrix satisfying Eq. (9.4) can be written as a scalar multiple of

$$\mathbf{J} = \begin{bmatrix} 0 & 1 \\ -1 & 0 \end{bmatrix}. \tag{9.5}$$

This means that infinitesimal rotations have the form $(x', y') \simeq (x + \theta y,\ y - \theta x)$, in agreement with Eq. (9.1).

Finite rotations are obtained by composing infinitely many infinitesimal ones, a process naturally described by the exponential map.

Proposition 9.1 *For* $\theta \in [0, 2\pi)$ *one has*

$$\mathbf{R}(\theta) = e^{\theta \mathbf{J}}. \tag{9.6}$$

Proof Given the multiplicative law of the rotation group, and the well known identity

$$e^x = \lim_{N \to \infty} \left(1 + \frac{x}{N}\right)^N$$

one has:

$$\mathbf{R}(\theta) = \lim_{n \to \infty} \left(\mathbf{R}\left(\frac{\theta}{n}\right)\right)^n = \lim_{n \to \infty} \left(\mathbf{I}_2 + \frac{\theta \mathbf{J}}{n}\right)^n = e^{\theta \mathbf{J}}.$$

□

It then follows that

$$\mathbf{J} = \left.\frac{d\mathbf{R}(\theta)}{d\theta}\right|_{\theta=0}. \tag{9.7}$$

Proposition 9.2 *Equation* (9.6) *implies Eq.* (9.1).

Proof Noting that $\mathbf{J}^2 = -\mathbf{I}_2$, which means that $\mathbf{J}$ plays in two dimensions the same role that the imaginary unit i plays in one dimension, we have

$$e^{\theta \mathbf{J}} = \sum_{k=0}^{\infty} \frac{\theta^k \mathbf{J}^k}{k!} = \sum_{k=0}^{\infty} (-1)^k \frac{\theta^{2k}}{(2k)!} \mathbf{I}_2 + \sum_{k=0}^{\infty} (-1)^k \frac{\theta^{2k+1}}{(2k+1)!} \mathbf{J} = \cos\theta\, \mathbf{I}_2 + \sin\theta\, \mathbf{J}.$$

□

One can observe that, using the Lie method, the expression for proper rotations is obtained without ever imposing $\det \mathbf{R} = 1$ explicitly. This follows from the fact that the Lie method is based on the dependence of the group on a continuous parameter, while O(n) consists of two disconnected components ($\det \mathbf{R} = 1$ and $\det \mathbf{R} = -1$), one of which is not continuously connected to the identity. Here $\mathbf{R}(\theta)$ is defined as continuously connected to $\mathbf{I}_2$.

Hence, we have recovered what we knew from trigonometry for a rotation of the coordinate axes (x, y) by an angle θ in two dimensions, because

$$\begin{aligned}\mathbf{R}(\theta) = e^{\theta\mathbf{J}} &= \left(\sum_{k=0}^{\infty}(-1)^k\frac{\theta^{2k}}{(2k)!}\right)\mathbf{I} + \left(\sum_{k=0}^{\infty}(-1)^k\frac{\theta^{2k+1}}{(2k+1)!}\right)\mathbf{J} \\ &= \cos\theta\,\mathbf{I} + \sin\theta\,\mathbf{J} \\ &= \cos\theta\begin{pmatrix}1 & 0\\ 0 & 1\end{pmatrix} + \sin\theta\begin{pmatrix}0 & 1\\ -1 & 0\end{pmatrix} \\ &= \begin{pmatrix}\cos\theta & \sin\theta\\ -\sin\theta & \cos\theta\end{pmatrix}\end{aligned} \tag{9.8}$$

which is the standard form of the rotation matrix in two dimensions.

9.1.2 Lie Algebra of Rotations

The generators introduced above form an algebraic structure that captures the essential features of the Lie group near the identity.

The Lie method is a general method that applies to all groups depending on a continuous parameter. In particular, it allows one to derive the expression for rotations in $\mathbb{R}^n$ with $n > 2$, for which a trigonometric approach is unfeasible. Indeed, the condition in Eq. (9.4) holds in any dimension.

In the case $n = 3$, the basic antisymmetric matrices are

$$J_x = \begin{bmatrix}0 & 0 & 0\\ 0 & 0 & 1\\ 0 & -1 & 0\end{bmatrix} \qquad J_y = \begin{bmatrix}0 & 0 & -1\\ 0 & 0 & 0\\ 1 & 0 & 0\end{bmatrix} \qquad J_z = \begin{bmatrix}0 & 1 & 0\\ -1 & 0 & 0\\ 0 & 0 & 0\end{bmatrix} \tag{9.9}$$

Three-dimensional rotations depend on three angles (again in $[0, 2\pi)$) and, near the identity, can be written as:

$$\mathbf{R}(\theta_x, \theta_y, \theta_z) = e^{\theta_x J_x + \theta_y J_y + \theta_z J_z}. \tag{9.10}$$

The matrices in Eq. (9.9) are real antisymmetric matrices. To obtain Hermitian generators (as commonly used in quantum mechanics), one defines

$$\mathrm{J}_k \equiv -i J_k. \tag{9.11}$$

In this way, rotations are expressed as

$$\mathbf{R}(\theta_x, \theta_y, \theta_z) = \exp\left[i \sum_{k=1}^{3} \theta_k \mathrm{J}_k \right]. \tag{9.12}$$

In higher dimensions, the non-commutativity of rotations carries essential geometric information. Unlike 2D rotations, rotations in $\mathbb{R}^n$ with $n > 2$ do not commute:

$$\mathbf{R}\mathbf{R}'\mathbf{R}^{-1} \simeq (\mathbf{I}_2+\mathbf{A})(\mathbf{I}_2+\mathbf{B})(\mathbf{I}_2-\mathbf{A}) = \mathbf{I}_2+\mathbf{B}+\mathbf{A}\mathbf{B}-\mathbf{B}\mathbf{A} \simeq \mathbf{R}'+[\mathbf{A}, \mathbf{B}]. \tag{9.13}$$

The commutation condition is therefore $[\mathbf{A}, \mathbf{B}] = 0$, where

$$[\mathbf{A}, \mathbf{B}] = -\sum_{i,j} \theta_i \theta_j [J_i, J_j], \tag{9.14}$$

where the commutator is defined as

$$[\mathbf{A}, \mathbf{B}] = \mathbf{A}\mathbf{B} - \mathbf{B}\mathbf{A}.$$

In the case $n = 3$, direct computation yields

$$[\mathrm{J}_i, \mathrm{J}_j] = i \sum_{k=1}^{3} \epsilon_{ijk} \mathrm{J}_k. \tag{9.15}$$

Formally, the J_k are called the *generators* of the *Lie group*, while their commutation relations define the *Lie algebra* of the group. Their mutual commutation relations define the associated Lie algebra.

This construction extends far beyond rotations and applies to any continuous symmetry group. In general, a Lie group may depend on an arbitrary set of k continuous parameters (such that $g(0, \dots, 0) = e$). The generators $\{T_j\}_{j=1,\dots,k}$ are defined through

$$g(\theta_1, \dots, \theta_k) = \exp\left[\sum_{\alpha=1}^{k} \theta_\alpha T_\alpha \right]. \tag{9.16}$$

The Lie algebra is determined by relations of the form

$$[T_\alpha, T_\beta] = \sum_{\gamma=1}^{k} f_{\alpha\beta\gamma} T_\gamma, \tag{9.17}$$

where the $f_{\alpha\beta\gamma}$ are called the *structure constants* of the algebra. For rotations in 3D, the structure constants coincide with the Levi-Civita symbol, cf. Eq. (9.15).

9.1.3 Rotations in n Dimensions

The Lie algebra approach allows rotations to be defined and analyzed in arbitrary dimensions, even when geometric intuition fails.

Rotations in $\mathbb{R}^n$ are described by SO(n), which depends on

$$\frac{1}{2}n(n-1)$$

independent parameters corresponding to the planes of rotation. These parameters can be labeled by pairs of indices

$$1 \leq a < b \leq n.$$

This number of generalized angles corresponds to the independent planes of rotation in n dimensions. The generators of SO(n) are therefore

$$(\mathrm{J}_{ab})_{ij} = -i(\delta_{ai}\delta_{bj} - \delta_{aj}\delta_{bi}) = \begin{cases} -i & i = a,\ j = b, \\ i & i = b,\ j = a, \\ 0 & \text{otherwise}, \end{cases} \qquad 1 \leq a < b \leq n. \tag{9.18}$$

and a generic rotation is

$$\mathbf{R}(\theta) = \exp\left[i \sum_{1 \leq a < b \leq n} \theta_{ab}\, \mathrm{J}_{ab} \right]. \tag{9.19}$$

Example

Verify that the group SO(n) satisfies all the group axioms.

First of all, we notice that the product of two rotations is a rotation,

$$(R_1 R_2)^T (R_1 R_2) = (R_2^T R_1^T)(R_1 R_2) \tag{9.20}$$

$$= R_2^T (R_1^T R_1) R_2 \tag{9.21}$$

$$= R_2^T R_2 \tag{9.22}$$

$$= I, \tag{9.23}$$

which implies that the first axiom is satisfied.

Moreover,

$$\det(R_1 R_2) = \det R_1 \ \det R_2 = 1.$$

Matrix multiplication is associative. The remaining two axioms are trivially satisfied.

Example
In $\mathbb{R}^3$, rotations about the z axis are generated by $J_{(12)}$, since they occur in the xy plane.

Proposition 9.3 *The Lie algebra of* SO(n) *is defined by*

$$[\mathrm{J}_{(ab)}, \mathrm{J}_{(pq)}] = i\left(\delta_{ap}\mathrm{J}_{(bq)} + \delta_{bq}\mathrm{J}_{(ap)} - \delta_{bp}\mathrm{J}_{(aq)} - \delta_{aq}\mathrm{J}_{(bp)}\right). \tag{9.24}$$

9.1.4 Generators in Differential Form

In physical applications, symmetry generators often act as differential operators on fields rather than as finite-dimensional matrices.

It is possible to express the generators of rotations as differential operators. For example, in $\mathbb{R}^2$,

$$\psi'(x', y') = \psi(x - \theta y,\ y + \theta x) \simeq \psi(x, y) + \theta\left(-y\frac{\partial}{\partial x} + x\frac{\partial}{\partial y}\right)\psi(x, y) \tag{9.25}$$

$$\simeq \left(1 + i\theta \hat{J}_z\right)\psi(x, y). \tag{9.26}$$

Where we used a Taylor expansion truncated to first order to represent

$$\psi(x', y') = \psi(x - y\theta,\ y + x\theta)$$

as

$$\psi(x + \varepsilon,\ y + \delta) = \psi(x, y) + \left(\frac{\partial \psi}{\partial x}\right)\varepsilon + \left(\frac{\partial \psi}{\partial y}\right)\delta + \text{h.o.t.}$$

with the identifications:

$$\varepsilon = -y\theta, \qquad \delta = x\theta.$$

$$\psi(x', y') = \psi(x, y) - \left[y\theta \frac{\partial}{\partial x} - x\theta \frac{\partial}{\partial y} \right] \psi(x, y).$$

$$\begin{pmatrix} x' \\ y' \end{pmatrix} = \left(\mathbf{I} + \theta \hat{J}_z \right) \begin{pmatrix} x \\ y \end{pmatrix}$$

which recovers:

$$\mathbf{R}(\theta) = \left(\mathbf{I} + \theta \hat{J}_z \right),$$

with

$$\hat{\mathrm{J}}_z = -i \, \hat{J}_z = -i \left(x \frac{\partial}{\partial y} - y \frac{\partial}{\partial x} \right).$$

And similar expressions can be obtained, similarly, for $\hat{\mathrm{J}}_x$ and $\hat{\mathrm{J}}_y$.

It is easy to verify the commutators that define the Lie algebra of SO(3) using the differential form of the rotation operators.

For example,

$$\left[y \frac{\partial}{\partial z}, z \frac{\partial}{\partial x} \right] = y \left(\frac{\partial z}{\partial z} \right) \frac{\partial}{\partial x} = y \frac{\partial}{\partial x}.$$

Thus,

$$[\hat{\mathrm{J}}_x, \hat{\mathrm{J}}_y] = \left[-i \left(y \frac{\partial}{\partial z} - z \frac{\partial}{\partial y} \right), -i \left(z \frac{\partial}{\partial x} - x \frac{\partial}{\partial z} \right) \right] \tag{9.27}$$

$$= i \, \hat{\mathrm{J}}_z. \tag{9.28}$$

The cyclic permutations then give

$$[\hat{\mathrm{J}}_i, \hat{\mathrm{J}}_j] = i \, \varepsilon_{ijk} \, \hat{\mathrm{J}}_k.$$

In the above calculations, the relation,

$$\left[\frac{\partial}{\partial z}, z \right] = 1.$$

has been used throughout.

Note the signs arising from the fact that $\hat{J}_z$ rotates the object and not the reference frame (transpose operation).

Generalizing to $\mathbb{R}^3$,

$$\hat{\mathbf{J}} = -i \, \mathbf{x} \times \nabla. \tag{9.29}$$

where $\mathbf{x} = (x, y, z)$. A direct calculation shows that these three generators satisfy Eq. (9.15). These are nothing but the angular momentum operators of quantum mechanics: $\mathbf{L} = \hbar\mathbf{J}$. This illustrates how Lie algebra generators acquire direct physical meaning as observables.

9.2 Lie Algebra Structure for a Generic Continuous Symmetry Group

The construction developed for rotations exemplifies a general pattern shared by all continuous symmetry groups. The general structure of a Lie algebra can be extracted systematically by analyzing group elements near the identity. The previous results are valid for any group whose elements $g(\theta_1, \theta_2, \dots)$ are continuously parametrized by a set of continuous parameters, and such that

$$g(0, 0, \dots) = I,$$

the identity element. For example, in SO(3) the continuous parameters are the angles θ_i, with $i = 1, 2, 3$.

These are therefore Lie groups. For them, the structure is always formed always in the following way:

1. Expand the elements of the group in the neighborhood of the identity:

$$g \simeq I + A.$$

2. Write

$$A = i \sum_{\alpha} \theta_\alpha T_\alpha,$$

   where the T_α are the generators.
3. Choose two elements A and B of the group, close to the identity:

$$g_1 \simeq I + A, \qquad g_2 \simeq I + B.$$

   Then

$$g_1 g_2 g_1^{-1} \simeq I + B + [A, B].$$

4. As in point (2), write

$$B = i \sum_{\beta} \theta'_\beta T_\beta.$$

In the same way, we write the commutator $[A, B]$ as a linear combination of the generators T_γ. By performing the substitutions, one obtains

$$[T_\alpha, T_\beta] = i \sum_\gamma f_{\alpha\beta\gamma} T_\gamma .$$

Again, the coefficients $f_{\alpha\beta\gamma}$ are called the *structure constants*.

In other words, the commutator of two generators can always be written as a linear combination of the generators themselves.

The above structure holds for any continuous group. The structure constants completely determine the Lie algebra of the specific Lie group.

Lie theory thus provides the essential link between continuous symmetries, their infinitesimal generators, and the algebraic structures that govern physical dynamics.

References

1. A. Zee, *Group Theory in a Nutshell for Physicists* (Princeton University Press, Princeton, 2016)
2. K.F. Riley, M.P. Hobson, S.J. Bence, *Mathematical Methods for Physics and Engineering*, 3rd edn. (Cambridge University Press, Cambridge, 2006)

Chapter 10
Introduction to Representation Theory

Abstract We shall now provide an informal introduction to the most important group homomorphism in physics, i.e. matrix representation. Each element of a group can be represented with a matrix, which is always true for rotations and also for other geometric transformations. We will start by introducing the basic concept of matrix representation and of equivalent representation. We shall then review the formal algebraic properties of representations and then introduce the concept of reducible and irreducible representations and how this can be formulated with the help of vector space theory. Again an excellent reference for this part is provided by Jones [1]. This chapter completes the transition from abstract group structure to concrete linear actions on physical state spaces, providing the language needed to analyze symmetry in classical and quantum systems.

10.1 Introduction to Matrix Representation

Having understood how continuous symmetries are generated infinitesimally through Lie algebras, we now ask how these symmetries act concretely on vectors and fields.

If F is a homomorphism from a group G to a group of non-singular $n \times n$ matrices, with matrix multiplication as the group operation, then the matrix group $F(G)$ forms a n-dimensional representation of the group G.

In physical terms, a representation specifies how symmetry transformations act on a space of states.

Definition 10.1 A representation $D : G \to \mathrm{GL}(V)$ is called *faithful* if it is injective, i.e. if

$$\ker(D) = \{e\},$$

where e is the identity element of the group G. In this case distinct elements of the group are represented by distinct linear transformations.

A. Zaccone, *Differential Geometry and Group Theory*,
https://doi.org/10.1007/978-3-032-27518-9_10

Faithful representations retain complete information about the group structure, while non-faithful ones identify distinct group elements as physically equivalent. **NB:** If e is the identity element of G, then

$$D(e) = I_d.$$

where I_d is the identity matrix of size $d \times d$. Importantly, for Lie groups the matrices must be continuous functions of the parameters $\theta_1, \theta_2, \ldots, \theta_n$.

10.2 Equivalent Representations

Different matrix realizations may describe the same group action when expressed in different bases. A pivotal concept is that of equivalent representations.

Two representations $D^{(1)}$ and $D^{(2)}$ are said to be equivalent,

$$D^{(1)} \sim D^{(2)},$$

if

$$D^{(1)}(g) = S\, D^{(2)}(g)\, S^{-1} \quad \forall\, g \in G,$$

where S is a non-singular matrix, independent of g.

Equivalent representations are, in essence, the same representation. This equivalence reflects the arbitrariness of the choice of basis in the representation space.

Furthermore, the equivalence transformation preserves the group structure. Indeed,

$$
\begin{aligned}
S\, D^{(2)}(g_1 g_2)\, S^{-1} &= S\, D^{(2)}(g_1) D^{(2)}(g_2)\, S^{-1} && (10.1)\\
&= \big(S D^{(2)}(g_1) S^{-1}\big)\big(S D^{(2)}(g_2) S^{-1}\big) && (10.2)\\
&= D^{(1)}(g_1)\, D^{(1)}(g_2) && (10.3)\\
&= D^{(1)}(g_1 g_2). && (10.4)
\end{aligned}
$$

The matrix S represents a change of basis in the representation space. Equivalent representations differ only by such a basis transformation and therefore describe the same group action, even though the matrices may look different. This notion is fundamental when classifying representations, since only non-equivalent representations carry genuinely new information.

10.3 Representation Algebra

To formalize these ideas, we now recast representations within the general framework of group homomorphisms. In this section we thus provide a more formal presentation of the algebraic properties of representations.

Definition 10.2 Given two groups $(A, *_A)$ and $(B, *_B)$, each equipped with its own group multiplication law ($*_A$ and $*_B$ respectively), a mapping $f : A \to B$ is called a *homomorphism* between A and B if

$$f(a_1 *_A a_2) = f(a_1) *_B f(a_2) \qquad \forall a_1, a_2 \in A.$$

Definition 10.3 Given a homomorphism $f : A \to B$, one defines its *image* or *range*

$$\operatorname{ran} f \equiv \{b \in B : b = f(a),\ a \in A\}$$

and its *kernel*

$$\ker f \equiv \{a \in A : f(a) = e_B\}.$$

Elements in the kernel act trivially in the representation and therefore encode redundancies in the group action.

Proposition 10.1 $\ker f \trianglelefteq A$.

Proof Given $a \in \ker f$, one has

$$f(gag^{-1}) = f(g)f(a)f(g^{-1}) = f(g)f(g^{-1}) = f(e_A) = e_B.$$

□

This observation anticipates the deep connection between representations and quotient groups.

Example
The map $f : D^4 \to \mathbb{Z}_2$ defined by

$$f(e) = f(c^2) = f(b) = f(bc^2) = +1, \quad f(c) = f(c^3) = f(bc) = f(bc^3) = -1$$

is a homomorphism because it is a surjective map but it is not injective. To be a isomorphism (see below) it must be both surjective and injective, hence bijective.

In practice, representations act on vector spaces through linear transformations.

Definition 10.4 Given a group G, a *linear representation* of G on a finite vector space $V(\mathbb{K})$ is a homomorphism

$$T : G \to \mathrm{GL}(V).$$

Proposition 10.2 $T(e) = \mathrm{id}_V$, *where* id_V *is the identity in the vector space* V.

In most applications, rather than working directly with linear maps $f \in \mathrm{GL}(V)$, one considers their associated matrices $M_f \in \mathrm{GL}(n, \mathbb{K})$, where $n = \dim_{\mathbb{K}} V$ (called the *degree* of the representation). Once a basis is chosen, group actions are encoded entirely in matrix form.

Definition 10.5 An *isomorphism* is a bijective homomorphism.

Example
Consider the determinant map

$$\det : \mathrm{O}(n) \to \{\pm 1\} \cong \mathbb{Z}_2.$$

This map is a group homomorphism, since

$$\det(AB) = \det(A)\det(B).$$

Its kernel is precisely the subgroup of orthogonal matrices with determinant $+1$, namely

$$\ker(\det) = \mathrm{SO}(n).$$

By the first isomorphism theorem,

$$\mathrm{O}(n)/\mathrm{SO}(n) \cong \mathbb{Z}_2.$$

This reflects the fact that orthogonal matrices have determinant ± 1, corresponding respectively to proper and improper rotations. Consequently $\mathrm{O}(n)$ can be viewed as an extension of $\mathrm{SO}(n)$ by $\mathbb{Z}_2$, and in general one writes

$$\mathrm{O}(n) \simeq \mathrm{SO}(n) \rtimes \mathbb{Z}_2.$$

Definition 10.6 Given a group G, two representations $D^{(1)}$, $D^{(2)}$ on vector spaces V, W are called *equivalent* (or *isomorphic* or *similar*) if there exists an isomorphism $\varphi : V \to W$ such that

$$\varphi \circ D^{(1)}(g) = D^{(2)}(g) \circ \varphi \quad \forall g \in G. \tag{10.5}$$

In matrix form,

$$D^{(1)} \sim D^{(2)} \;\Leftrightarrow\; \exists S \in \mathrm{GL}(n, \mathbb{K}) :\; D^{(2)}(g) = S D^{(1)}(g) S^{-1} \quad \forall g \in G.$$

It is immediate to see that similarity preserves the group structure:

$$D^{(2)}(g_1g_2)=SD^{(1)}(g_1g_2)S^{-1}=SD^{(1)}(g_1)S^{-1}SD^{(1)}(g_2)S^{-1} = D^{(2)}(g_1)D^{(2)}(g_2).$$

Definition 10.7 Given a group G, a *unitary representation* is a representation

$$D : G \to \mathrm{U}(n).$$

where $\mathrm{U}(n)$ denotes the group of $n \times n$ unitary matrices.

In other words, a representation is unitary if every group element is represented by a unitary matrix, so that the action of the group preserves the inner product structure of the vector space. In physics this property is essential, since it ensures the conservation of probability amplitudes and norms of state vectors.

When continuity enters the picture, representations must respect the underlying topological structure of the group. A topological structure can be assigned to a group.

Definition 10.8 A *topological group* G is a group endowed with a Hausdorff topological space structure such that the maps $(a, b) \mapsto ab$ and $a \mapsto a^{-1}$ are continuous.

Definition 10.9 A topological group G is called a *Lie group* if the maps $(a, b) \mapsto ab$ and $a \mapsto a^{-1}$ are smooth.

Informally, Lie groups can be viewed both as groups and as differentiable manifolds (more precisely, there exists a diffeomorphism between the two structures).

The interplay between topology and representation theory becomes especially transparent for rotation groups, as we shall see in the following examples.

Examples

- SO(2) corresponds to $\mathbb{S}^1$.
- SO(3) corresponds to $\mathbb{D}^3/\sim$, where $\sim$ is the equivalence relation that associates to each point its antipodal point.

  In the above examples, the space $\mathbb{S}^1$ denotes the unit circle in the plane,

$$\mathbb{S}^1 = \{(x, y) \in \mathbb{R}^2 \mid x^2 + y^2 = 1\}.$$

The group SO(2) consists of all rotations in two dimensions. Each element of SO(2) is completely determined by a rotation angle θ, with angles that differ by 2π describing the same rotation. This establishes a one-to-one correspondence between rotations and points on the unit circle via

$$\theta \longleftrightarrow (\cos\theta, \sin\theta),$$

so that, both topologically and geometrically,

$$\mathrm{SO}(2) \cong \mathbb{S}^1.$$

The space $\mathbb{D}^3$ denotes the closed unit ball in $\mathbb{R}^3$,

$$\mathbb{D}^3 = \{\mathbf{x} \in \mathbb{R}^3 \mid \|\mathbf{x}\| \leq 1\}.$$

The quotient $\mathbb{D}^3/\sim$ is obtained by identifying antipodal points on the boundary of the ball, that is, by imposing the equivalence relation

$$\mathbf{x} \sim -\mathbf{x} \qquad \text{for all } \|\mathbf{x}\| = 1.$$

Points in the interior of the ball are left unchanged, while each pair of opposite points on the surface $\partial\mathbb{D}^3 = \mathbb{S}^2$ is identified as a single point.

This construction provides a geometric description of the group SO(3). Any rotation in three dimensions can be represented by an axis $\hat{\mathbf{n}}$ and an angle $\theta \in [0, \pi]$. The pair $(\theta, \hat{\mathbf{n}})$ and $(\theta, -\hat{\mathbf{n}})$ describe the same rotation when $\theta = \pi$, which corresponds precisely to the antipodal identification on the boundary of the ball. Consequently, the space of all rotations in three dimensions is topologically equivalent to the quotient space

$$\mathrm{SO}(3) \cong \mathbb{D}^3/\sim.$$

Once a topological structure is given, one can discuss the compactness of a group; see, for example, Definition 7.5. In the case of Lie groups, compactness can be determined by studying the associated differentiable manifold.

Proposition 10.3 *For compact Lie groups, every representation is equivalent to a unitary representation.*

10.4 Reducibility

A central question in representation theory is whether a representation can be decomposed into simpler, independent components. To address this question, we shall now informally introduce the pivotal concept of reducible (and irreducible) representations. Let

$$D(g) = \begin{pmatrix} A(g) & C(g) \\ 0 & B(g) \end{pmatrix},$$

with

$$A(g) \in \mathbb{C}^{m\times m}, \qquad B(g) \in \mathbb{C}^{n\times n}, \qquad C(g) \in \mathbb{C}^{m\times n}.$$

Then

$$D(g)\,D(g') = \begin{pmatrix} A(g) & C(g) \\ 0 & B(g) \end{pmatrix} \begin{pmatrix} A(g') & C(g') \\ 0 & B(g') \end{pmatrix} \tag{10.6}$$

$$= \begin{pmatrix} A(g)A(g') & A(g)C(g') + C(g)B(g') \\ 0 & B(g)B(g') \end{pmatrix}. \tag{10.7}$$

It follows that

$$A(gg') = A(g)A(g'), \qquad B(gg') = B(g)B(g').$$

Therefore $A(g)$ and $B(g)$ are homomorphisms and preserve the group composition law. Hence,

$A(g)$ is an m-dimensional representation,

$B(g)$ is an n-dimensional representation.

The matrix $C(g)$ does not define a representation, since it does not satisfy a closed composition law by itself. Its presence signals a coupling between subspaces that prevents a complete decomposition.

If $C(g) = 0$, the representation takes the block-diagonal form

$$D(g) = A(g) \oplus B(g).$$

In this case, the representation space splits into two invariant subspaces of dimensions m and n. The group action does not mix these subspaces, and the representation decomposes into a direct sum of two smaller representations. For this reason, the representation $D(g)$ is said to be *reducible*. Physically, reducibility means that the symmetry acts independently on distinct sectors of the state space.

Let

$$\mathbf{v} = x\hat{\imath} + y\hat{\jmath} + z\hat{k}$$

be a vector in three dimensions, where $\hat{\imath}$, $\hat{\jmath}$ and $\hat{k}$ are unit vectors in the three cartesian directions.

We decompose the vector space as

$$\mathbf{v} = \mathbf{u} + \mathbf{w}, \qquad \mathbf{u} = x\hat{\imath} + y\hat{\jmath}, \qquad \mathbf{w} = z\hat{k}.$$

Thus, the representation naturally decomposes into two separate representations: one of dimension 2 acting on $\mathbf{u}$, and one of dimension 1 acting on $\mathbf{w}$. These two representations are completely independent.

Indeed, from simple linear algebra we know that:

$$RR^T = \begin{pmatrix} A & 0 \\ 0 & 1 \end{pmatrix} \begin{pmatrix} A^T & 0 \\ 0 & 1 \end{pmatrix} = \begin{pmatrix} AA^T & 0 \\ 0 & 1 \end{pmatrix}.$$

This block decomposition is therefore stable and can be written as

$$R(g) = D^{(1)}(g) \oplus D^{(2)}(g).$$

In general, a reducible representation can be written as

$$D(g) = \begin{pmatrix} A(g) & C(g) \\ 0 & B(g) \end{pmatrix}, \qquad \forall g \in G.$$

For finite groups (and, in fact, also for compact Lie groups as we shall see), the matrix $C(g)$ can always be reduced to zero by an equivalence transformation,

$$SC(g)S^{-1} = 0.$$

In this case, the representation is said to be *completely reducible*, and

$$D(g) = A(g) \oplus B(g).$$

The representations A and B are then called the *irreducible representations*.

A proper subspace W of a vector space V is a subspace of smaller dimension (but not zero). A subspace W is said to be invariant under the action of a representation $D(g)$ if $D(g)$ maps every vector $v \in W$ into a vector $D(g)v \in W$.

A representation that has a proper invariant subspace is called reducible.

10.5 Character of a Representation

To classify representations without reference to a specific basis, one introduces basis-independent quantities.

The *character* of a representation D is the function

$$\chi : G \to \mathbb{C}, \qquad \chi(g) = \mathrm{Tr}[D(g)].$$

NB: The character does not distinguish between equivalent representations, but it does distinguish non-equivalent ones: The character is a function of the (equivalence) class. In other words, the character depends only on the conjugacy class of the group element.

Indeed, the trace of a product of matrices is cyclic:

$$
\begin{aligned}
\mathrm{Tr}(ABC) &= \sum_{ijk} A_{ij} B_{jk} C_{ki} \\
&= \sum_{ijk} B_{jk} C_{ki} A_{ij} \\
&= \mathrm{Tr}(BCA).
\end{aligned} \tag{10.8}
$$

And therefore:

$$
\begin{aligned}
\mathrm{Tr}\Big[S\,D(g)\,S^{-1}\Big] &= \sum_{i,k,l} S_{ik}\,D_{kl}(g)\,S^{-1}_{li} \\
&= \sum_{k,l} \left(S^{-1}S\right)_{lk} D_{kl}(g) \\
&= \sum_{k} D_{kk}(g) \\
&= \mathrm{Tr}[D(g)].
\end{aligned} \tag{10.9}
$$

Hence, if

$$
D^{(1)}(g) \sim D^{(2)}(g),
$$

then, also

$$
\{\chi^{(1)}(g)\} = \{\chi^{(2)}(g)\}.
$$

10.6 Connection with the Theory of Vector Spaces

The structure of representations mirrors familiar concepts from linear algebra. In this section we shall clarify that equivalent representations arise from changes of basis, just as in ordinary linear algebra. Reducibility appears when the vector space contains a subspace invariant under all group transformations. In such a basis, the representation matrices acquire a block-triangular (or block-diagonal) form, making the invariant structure explicit.

10.6.1 Similarity Transformation and Change of Basis

Let

$$
\mathbf{u} = u_i\,\mathbf{e}_i
$$

be a vector written in the basis $\{\mathbf{e_i}\}$. Under a linear transformation T, represented by matrix D_{ij} in the chosen basis $\{\mathbf{e_i}\}$,

$$v_i = D_{ij}\, u_j, \qquad T\mathbf{e}_j = D_{ij}\, \mathbf{e}_i, \qquad T\mathbf{u} = \mathbf{v},$$

hence

$$v_i = D_{ij} u_j.$$

If we change basis to $\{\mathbf{e}'_i\}$, related by

$$\mathbf{e}'_i = S_{ij}\mathbf{e}_j,$$

then

$$\mathbf{u} = u'_i\, \mathbf{e}'_i.$$

Furthermore, in the new basis we have:

$$v'_i = D'_{ij}\, u'_j, \qquad T\mathbf{e}'_j = D'_{ij}\, \mathbf{e}'_i.$$

In this basis the linear transformation is described by a different matrix, in general

$$D'_{ij} \neq D_{ij}.$$

The two bases are related by a change of basis:

$$\mathbf{e}_i = S_{ji}\, \mathbf{e}'_j, \qquad \mathbf{u} = u_i \mathbf{e}_i = u_i S_{ji} \mathbf{e}'_j.$$

and therefore, also:

$$u' = Su, \qquad v' = Sv.$$

$$v' = SDu = SDS^{-1}u' \quad \Rightarrow \quad D' = SDS^{-1}.$$

Equivalently, $D = S^{-1}D'S$. Thus, matrices representing the same linear transformation (T, in this example) but written in different bases are related by similarity transformations. This is precisely the notion of equivalence introduced earlier for group representations.

10.6.2 Representations of a Group and Linear Transformations

Matrices that represent the same linear transformation but using different bases are related by a similarity transformation; the conjugating matrix S is the one that relates the two bases.

We can view a representation of a group G as a homomorphism into the group of linear transformations T acting on a vector space V:

$$T(gg') = T(g)\,T(g').$$

A set of linear operators $\{T(g)\}$ corresponds to an entire equivalence class of representations, hence

$$D'(g) = SD(g)S^{-1}.$$

10.6.3 Reducibility Revisited

Reducibility corresponds to the existence of a subspace $U \subset V$ which is closed under the action of the group:

$$\mathbf{u} \in U \subset V \;\Rightarrow\; T(g)\mathbf{u} \in U.$$

Let $\{\mathbf{e}_i\}$, $i = 1, \ldots, m$, be a basis of U. We can extend it with vectors $\{\mathbf{e}_i\}$, $i = m+1, \ldots, m+n$, to form a basis of V (of dimension $m+n$). In this basis the matrix $D(g)$ corresponding to the transformation $T(g)$ is defined by

$$T(g)\mathbf{e}_j = D_{ij}(g)\,\mathbf{e}_i\,.$$

If W is the space spanned by $\{\mathbf{e}_i\}$ with $i = m+1, \ldots, m+n$, then the closure of U (i.e. the fact that U has no components in W) is reflected in the statement that

$$D_{ij}(g) = 0 \quad \text{for } j = 1, \ldots, m \text{ and } i = m+1, \ldots, m+n.$$

Therefore

$$D(g) = \begin{pmatrix} A(g) & C(g) \\ 0 & B(g) \end{pmatrix}.$$

Example: Group C_3
Let us consider the generic form of an element of C_3:

$$D_{ij} = \begin{pmatrix} A & B & 0 \\ C & D & 0 \\ 0 & 0 & 1 \end{pmatrix}$$

Then we can write:

$$T(g)\mathbf{e}_j = D_{ij}(g)\,\mathbf{e}_i, \qquad \mathbf{e}_i = \begin{pmatrix} x \\ y \\ 0 \end{pmatrix} \in U.$$

$$\begin{pmatrix} x & y & 0 \end{pmatrix} \begin{pmatrix} A & B & 0 \\ C & D & 0 \\ 0 & 0 & 1 \end{pmatrix} = \begin{pmatrix} Ax + Cy \\ Bx + Dy \\ 0 \end{pmatrix} \in U.$$

We thus obtain a vector that is still in U because it is a linear combination of x and y. If the last component were not 0, we would obtain a vector that is a linear combination of x, y, z and therefore it would not belong to U. This fact illustrates the link between the matrix block structure and the decomposition of the vector space associated with the representation (V) into mutually orthogonal invariant subspaces. In other words, the block (upper-triangular or block-diagonal) structure of the matrices of a reducible representation is the algebraic manifestation of the existence of a non-trivial invariant subspace, since in an adapted basis the group action does not mix vectors inside the invariant subspace with vectors outside it. The invariant subspace is sometimes called submodule or G-module.

10.7 Unitary Representations and Unitarity Theorem

In physics, representations are required to preserve probability amplitudes and inner products. This is, indeed, one of the most basic notions of "symmetry" in physics.

10.7.1 Unitary Transformations

A transformation T is unitary if it preserves the scalar product:

$$(T\mathbf{u}, T\mathbf{v}) = (\mathbf{u}, \mathbf{v}) \qquad \forall\, \mathbf{u}, \mathbf{v} \in V.$$

By definition, the operators $T(g)$ are invertible. Indeed, we can apply the operator T^{-1}, which also leaves the scalar product unchanged:

$$(T\mathbf{u}, \mathbf{v}) = (\mathbf{u}, T^{-1}\mathbf{v}).$$

This implies, in matrix form,

$$D^*_{ji}\, u_j\, v_i = u^*_j\, D^{-1}_{ji}\, v_i.$$

Therefore,

$$D^\dagger = D^{-1},$$

that is, the matrices D_{ij} which represent the operator T in a chosen basis, are unitary matrices. Unitarity therefore imposes strong constraints on the possible structure of a representation.

We have seen that for reducible matrices the vector space associated with the linear transformation $T(g)$ decomposes in such a way that

$$\mathbf{u} \in U \;\Rightarrow\; T(g)\mathbf{u} \in U.$$

The complement of U with respect to the full vector space V is a subspace W.

Having defined a scalar product, we can choose the basis of W in such a way that it is orthonormal to the basis of U.

Then,

$$(T(g)\mathbf{w}, \mathbf{u}) = (\mathbf{w}, T^{-1}(g)\mathbf{u}), \qquad \text{with } T \text{ unitary}.$$

Since

$$T^{-1}(g)\mathbf{u} = T(g^{-1})\mathbf{u} \in U,$$

and $\mathbf{w} \in W$, we obtain

$$(T(g)\mathbf{w}, \mathbf{u}) = (\mathbf{w}, \mathbf{u}) = 0.$$

Thus,

$$\Rightarrow \quad T(g)\mathbf{w} = \mathbf{w}' \in W$$

and it has no component in U. Hence,

$$W = \{\mathbf{w} \in V \mid (\mathbf{w}, \mathbf{u}) = 0 \;\; \forall \mathbf{u} \in U\},$$

that is, W is the orthogonal complement of U. That is to say, in a unitary representation, the existence of an invariant subspace U automatically implies that its orthogonal complement W is also invariant, which guarantees complete reducibility as we shall now see.

This implies that

$$C(g) = 0.$$

Therefore,

$$D(g) = \begin{pmatrix} A(g) & 0 \\ 0 & B(g) \end{pmatrix},$$

where the upper block acts on the subspace U and the lower block acts on the subspace W, *independently* of each other.

Proposition 10.4 *Unitary representations are completely reducible.*

Indeed, reducibility implies the existence of an invariant subspace U, and unitarity guarantees that the orthogonal complement W is also invariant. Consequently, the representation decomposes into a direct sum of irreducible representations. According to a theorem due to Maschke, for finite groups it is always possible to define and choose a suitable scalar product, which renders T a unitary operator. This result guarantees that symmetry analysis can always be reduced to irreducible building blocks.

10.7.2 Unitarity Theorem

All finite groups (in particular, all compact groups) admit unitary representations.

That is,

$$D^{\dagger}(g)\, D(g) = I \qquad \forall g \in G.$$

where I is the identity matrix.

Consequently, all reducible representations of finite (compact) groups are completely reducible.

10.7.3 Example: Group U(1)

In lieu of a formal demonstration, we illustrate the unitarity theorem with an example [1]. Suppose the representation $D(g)$ is 1×1, i.e. it is represented by a complex number:

$$D(g) = z \in \mathbb{C}.$$

For a finite group, for every element g of a finite group there exists an integer $k > 0$ such that

$$g^k = I,$$

that is, each element has finite order.

The element g^k is represented by

$$D(g^k) = [D(g)]^k,$$

so that

$$z^k = e^{ik\theta}.$$

If $n = 1$, then

$$D^\dagger(g)\, D(g) = e^{-i\theta} e^{i\theta} = 1,$$

and therefore

$$D(g) \text{ is unitary.}$$

That is to say, for finite (or compact) groups, unitarity can always be imposed without loss of generality, and this guarantees that any reducible representation decomposes into a direct sum of irreducible unitary representations.

The theorem extends from finite groups to compact Lie groups. For compact groups the averaging procedure used in the proof can be performed using the invariant (Haar) measure on the group. Compact Lie groups possess a finite invariant (Haar) measure. This allows averages over the group to be defined, which is the key ingredient in proving the unitarity theorem.

For continuous groups, the discrete sum over group elements is replaced by an integral:

$$\sum_g \longrightarrow \int d\mu(g),$$

where $d\mu(g)$ is the invariant (Haar) measure. The existence of such a measure is a direct consequence of compactness.

Rotation groups are parametrized by angles

$$\theta \in [0, 2\pi],$$

and for this reason the integral always converges. In other words, compactness guarantees the existence of a finite, invariant (Haar) measure, allowing the averaging procedure used in the unitarity theorem to remain valid even for infinite continuous groups such as rotation groups.

10.8 Exercises

1. Construct the 3×3 representation of the group D_3 in the basis (x, y, z). Compute the characters of all the elements of the group. Verify that the characters are the same for all the elements of the same conjugacy class, and explain why.
2. Verify that

$$D(c) = \begin{pmatrix} 0 & 1 \\ -1 & -1 \end{pmatrix}$$

generates a 2×2 representation of C_3. Show that the representation is irreducible over $\mathbb{R}$.

Solutions can be found in the end matter of [2].

Representation theory thus provides the final link between abstract symmetry groups and their concrete action on physical states, forming the backbone of modern quantum theory and field theory.

References

1. A. Zee, *Group Theory in a Nutshell for Physicists* (Princeton University Press, Princeton, 2016)
2. H.F. Jones, *Groups, Representations and Physics*, 2nd edn. (Taylor & Francis Ltd, Bristol/New York, 1998)

Chapter 11
Schur's Lemmas, the Great Orthogonality Theorem and the Regular Representation

Abstract In this chapter we delve deeper into the conceptual structure of representation theory, by presenting the key formal results upon which the whole theory rests upon. These key results are the Schur's lemmas, which, in turn, are essential to prove the great orthogonality theorem. The latter eventually provides the practical tools that are needed to construct the character tables of groups. These tools are the orthogonality relations for characters. Together, these results reveal the deep connection between group structure, irreducible representations, and conjugacy classes, culminating in the regular representation as a unifying object.

11.1 Schur's Lemmas

In the previous chapter we saw that unitary representations of finite or compact groups decompose into irreducible components. We now investigate the algebraic rigidity of irreducible representations themselves.

We have seen that, for finite or compact groups, the representations are either irreducible or completely reducible. The matrices of a reducible representation can be brought into block-diagonal form, where the matrices along the diagonal correspond to irreducible representations (irreps). The irreducible representations therefore play a fundamental role, as formalized by Schur's lemmas. The latter characterize all linear operators that commute with the action of the group on an irreducible representation.

11.1.1 The First Schur Lemma

Schur's first lemma states that irreducibility leaves no room for nontrivial operators commuting with the group action: the only such operators are scalar multiples of the identity.

In plain words the first lemma says: Every matrix that commutes with all matrices of an irreducible representation must be proportional to the identity matrix.

A. Zaccone, *Differential Geometry and Group Theory*,
https://doi.org/10.1007/978-3-032-27518-9_11

This statement can be expressed compactly in algebraic form. That is,

$$B\, D(g) = D(g)\, B \qquad \forall g \in G$$

implies

$$B = \lambda\, I.$$

In terms of linear operators,

$$\hat{B}\, T(g) = T(g)\, \hat{B} \qquad \forall g \in G \quad \Rightarrow \quad \hat{B} = \lambda\, \hat{E}.$$

Here $\hat{B} : V \to V$ is a linear operator corresponding to the matrix B, and $\hat{E}$ is the identity operator on the vector space V, defined by

$$\hat{E}\, \mathbf{u} = \mathbf{u}.$$

Proof
Let $\mathbf{b}$ be an eigenvector of $\hat{B}$ with eigenvalue λ:

$$\hat{B}\mathbf{b} = \lambda\mathbf{b}.$$

Then

$$\hat{B}\big(T(g)\mathbf{b}\big) = T(g)\hat{B}\mathbf{b} = T(g)(\lambda\mathbf{b}) = \lambda\, T(g)\mathbf{b}.$$

Therefore $T(g)\mathbf{b}$ is also an eigenvector of $\hat{B}$ with the same eigenvalue λ.

The eigenvectors of $\hat{B}$ with eigenvalue λ form a vector subspace U, which is closed under the action of the group through the operators $\{T(g)\}$:

$$T(g_1)\big(T(g_2)\mathbf{b}\big) = T(g_1 g_2)\mathbf{b} \in U.$$

The key observation is that commutation with the group action forces this subspace U to be invariant.

It is closed because $T(g)\mathbf{b}$ is eigenvector of $\hat{B}$ with the same eigenvalue as $\mathbf{b}$ and hence still belongs to U. Now we invoke *irreducibility*. The subspace U cannot be decomposed further: the only invariant subspaces are $U = V$ or $U = \{0\}$.

Since we work over $\mathbb{C}$, the second case is impossible, because the characteristic polynomial of B has at least one root by the fundamental theorem of algebra. Hence B admits at least one eigenvector. Indeed, the characteristic equation

$$\det(B - \lambda I) = 0$$

always admits a non-zero solution λ.

Hence the space of eigenvectors of $\hat{B}$ with eigenvalue λ is the entire space:

$$U = V.$$

Therefore,

$$\hat{B}\mathbf{u} = \lambda\mathbf{u} \qquad \forall\, \mathbf{u} \in V,$$

which implies

$$\hat{B} = \lambda\hat{E}.$$

This proves the first lemma. Thus, irreducibility enforces maximal simplicity: only scalar operators commute with the group action.

11.1.2 The Second Schur Lemma

While the first Schur lemma concerns operators acting within a single irreducible representation, the second lemma addresses maps between different irreducible representations. Schur's second lemma states that no non-trivial inter-twiner exists between inequivalent irreducible representations: the only operator commuting with the group action is the zero map. Of this lemma we only provide the statement. The proof can be found in [1] or [2].

Let $D(g)$ and $D'(g)$ be irreducible representations.

If

$$B\,D(g) = D'(g)\,B \qquad \forall g \in G,$$

then either $B = 0$ or the two representations are equivalent. In the latter case one has

$$B = \lambda I.$$

The corresponding linear operator $\hat{B}$ is a mapping

$$\hat{B} : V \longrightarrow V',$$

between two vector spaces V and V' that support the two representations, and satisfies

$$\hat{B}\,T(g) = T'(g)\,\hat{B} \qquad \forall g \in G \quad \Rightarrow \quad \hat{B} = \hat{0}.$$

Here $\hat{0}$ is the linear operator that maps every vector in V to the zero vector in V':

$$\hat{0}\,\mathbf{u} = \mathbf{0} \qquad \forall \mathbf{u} \in V.$$

11.2 The Great Orthogonality Theorem

Schur's lemmas provide the algebraic backbone for one of the most powerful results in representation theory: the orthogonality relations for irreducible representations.

Let U_1 and U_2 be vector subspaces (i.e. G-modules) associated with irreducible, non-equivalent representations. Suppose there exists a linear map

$$A : U_1 \longrightarrow U_2$$

such that

$$T^{(1)}(g)\,A = A\,T^{(2)}(g).$$

To exploit Schur's lemmas, we build an operator by averaging over the group action. We then construct the operator

$$\hat{B} = \frac{1}{|G|}\sum_{g\in G} T^{(1)}(g)\,\hat{A}\,T^{(2)}(g)^{-1}.$$

At this point, we should set up the key step: $\hat{B}$ commutes with the entire irreducible representation $T^{(1)}$, so by Schur's second lemma it must vanish—which leads directly to orthogonality relations between matrix elements of inequivalent irreps.

Here the sum runs over the elements of the group. For compact groups, instead of the discrete sum we use an integral $\int ...d\mu(g)$.

Let $h \in G$. Then

$$T^{(1)}(h)\,\hat{B} = \sum_{g} T^{(1)}(h)\,T^{(1)}(g)\,\hat{A}\,T^{(2)}(g)^{-1}.$$

Using the group property of the representation, we relabel $g \rightarrow hg$ in the argument of $T^{(1)}$; correspondingly, the argument of $T^{(2)}$ becomes $(hg)^{-1}$. Moreover, since $\sum_g = \sum_{hg}$, we obtain

$$T^{(1)}(h)\,\hat{B} = \sum_{g} T^{(1)}(g)\,\hat{A}\,T^{(2)}(g)^{-1} = \hat{B}.$$

Using the group property, the argument of $T^{(1)}$ becomes $(g^{-1}h)$. Thus,

$$T^{(1)}(h)\,\hat{B} = \sum_g T^{(1)}(g)\,\hat{A}\,T^{(2)}(g^{-1})\,T^{(2)}(h).$$

Recognizing the sum, we obtain

$$T^{(1)}(h)\,\hat{B} = \hat{B}\,T^{(2)}(h).$$

We have therefore shown that $\hat{B}$ satisfies the hypotheses of Schur's lemma, and hence

$$\hat{B} = \hat{0},$$

unless $U_1 = U_2$. This vanishing condition translates directly into orthogonality relations between matrix elements.

In this latter case the two representations coincide, and by the first Schur lemma we have

$$\hat{B} = \lambda\,\hat{E}.$$

Therefore,

$$\hat{B} = \sum_g T^{(\mu)}(g)\,\hat{A}\,T^{(\mu)}(g)^{-1} = \lambda_A^{(\mu)}\,\hat{E}.$$

The constant λ depends both on the index μ of the irreducible representation and on the choice of the operator A.

Choosing A such that its only nonzero matrix element is

$$(A)_{mn} = \delta_{ms}\delta_{nt},$$

we obtain

$$\lambda_A^{(\mu)} = \lambda_{st}^{(\mu)}.$$

and, thus,

$$\sum_g D_{rs}^{(\mu)}(g)\,D_{tq}^{(\nu)}(g^{-1}) = \lambda_{rs}^{(\mu)}\,\delta^{\mu\nu}\,\delta_{rq}.$$

For $\mu = \nu$, we contract the indices r and q, i.e. we take the trace:

$$\sum_g D_{rs}^{(\mu)}(g)\,D_{sr}^{(\mu)}(g^{-1}) = \sum_g \left(D^{(\mu)}(g^{-1})D^{(\mu)}(g)\right)_{ss} = \sum_g n_\mu\,\lambda_{rs}^{(\mu)}.$$

Here n_μ is the dimension of the irreducible representation $D^{(\mu)}$.

Since

$$D^{(\mu)}(g^{-1})D^{(\mu)}(g) = I,$$

and the trace of the identity matrix of dimension n_μ is n_μ, we find

$$\sum_g \delta_{rs} = |G|\,\delta_{rs}.$$

Therefore,

$$\lambda_{rs}^{(\mu)} = \frac{|G|}{n_\mu}\,\delta_{rs}.$$

Substituting, we obtain the fundamental orthogonality relation for the matrices of irreducible representations:

$$\sum_g D_{ir}^{(\mu)}(g)\,D_{js}^{(\nu)}(g^{-1}) = \frac{|G|}{n_\mu}\,\delta^{\mu\nu}\,\delta_{ij}\,\delta_{rs}.$$

If the representations are unitary, as is the case for finite or compact groups, then

$$\sum_g D_{ri}^{(\mu)}(g)\,D_{sj}^{(\nu)*}(g) = \frac{|G|}{n_\mu}\,\delta^{\mu\nu}\,\delta_{ij}\,\delta_{rs}. \tag{11.1}$$

Let us now set $\mu = \nu$ and fix the indices i and r. Then the set

$$\left\{D_{ir}^{(\mu)}(g_1),\ \ldots,\ D_{ir}^{(\mu)}\left(g_{|G|}\right)\right\}$$

can be regarded as column vectors of dimension $|G|$. This interpretation makes orthogonality geometrically transparent. Then the sum on the left-hand side of Eq. (11.1) can be interpreted as a scalar product between two vectors in a vector space of dimension $|G|$. The indices i and s label the vector components, while the group element g labels the entries of the vector.

Each pair of indices (i, s) defines a vector. Since $i, s = 1, \ldots, n_\mu$, there are n_μ^2 such vectors, and they are all mutually orthogonal.

Considering all irreducible representations (indexed by μ), we obtain in total

$$\sum_\mu n_\mu^2$$

mutually orthogonal vectors.

The number of mutually orthogonal vectors cannot exceed the dimension of the corresponding vector space. Since these vectors belong to a space of dimension $|G|$, it follows that

$$\sum_{\mu} n_{\mu}^2 \leq |G|.$$

This inequality already hints at a deep constraint relating group order and irreducible representations.

11.3 Orthogonality of Characters

By taking traces of the matrix-element orthogonality relations, we now obtain corresponding scalar relations for characters. These are obtained by summing the matrix-element orthogonality relations over the diagonal indices, i.e. by taking traces. Since characters are class functions (they depend only on the conjugacy class of a group element), this collapses the detailed matrix structure into scalar relations that classify irreducible representations.

We recall from the previous chapter that the character of a representation D is the function

$$\chi : G \to \mathbb{C}, \qquad \chi(g) = \operatorname{Tr} D(g).$$

11.3.1 Properties of the Trace

1. As demonstrated in Chap. 10, the character is the same for equivalent representations:

$$D'(g) = SD(g)S^{-1}.$$

2. The character is the same for conjugate elements of the group:

$$D(hgh^{-1}) = D(h)\, D(g)\, D(h)^{-1}.$$

3. If the representation D is unitary, i.e.

$$D^{-1}(g) = D^{\dagger}(g),$$

then

$$\chi(g^{-1}) = \operatorname{Tr}\left(D(g)^{\dagger}\right) = \operatorname{Tr}(D(g))^{*} = \chi^{*}(g).$$

Using these properties and the orthogonality of matrix elements, we obtain

$$\sum_g \chi^{(\mu)}(g)\, \chi^{(\nu)}(g^{-1}) = \frac{|G|}{n_\mu}\, \delta_{\mu\nu}.$$

Here $\delta_{ij}\delta_{ji} = \delta_{ii} = n_\mu$. The summation convention is always understood, with indices running from 1 to n_μ.

Thus,

$$\frac{1}{|G|} \sum_g \chi^{(\mu)}(g)\, \chi^{(\nu)}(g^{-1}) = \delta^{\mu\nu}.$$

Using property (iii), we obtain

$$\frac{1}{|G|} \sum_g \chi^{(\mu)}(g)\, \chi^{(\nu)*}(g) = \delta^{\mu\nu}.$$

This is the scalar product of two column vectors of dimension $|G|$:

$$\left\{\chi^{(\mu)}(g_1), \chi^{(\mu)}(g_2), \dots, \chi^{(\mu)}(g_{|G|})\right\}$$

and

$$\left\{\chi^{(\nu)}(g_1), \chi^{(\nu)}(g_2), \dots, \chi^{(\nu)}(g_{|G|})\right\}.$$

It is therefore convenient to define the scalar product of two characters φ and χ as

$$\langle \varphi, \chi \rangle = \frac{1}{|G|} \sum_g \varphi(g)\, \chi(g^{-1}) = \langle \chi, \varphi \rangle.$$

Hence, the characters of irreducible representations are orthonormal:

$$\langle \chi^{(\mu)}, \chi^{(\nu)} \rangle = \delta^{\mu\nu}.$$

This is the final form of character orthogonality: irreducible characters form an orthonormal basis in the space of class functions on the group. This result underlies decomposition formulas, projection operators, and the character table machinery that we shall study in the next chapter.

By virtue of property (ii), all elements belonging to the same conjugacy class have the same character. Therefore, distinct characters can be labeled as follows.

Let

$$\chi_i, \quad i = 1, \dots, k,$$

be the characters in correspondence with the k conjugacy classes.

Let p_i be the number of elements in the i-th conjugacy class. Then the sum

$$\frac{1}{|G|}\sum_g \chi^{(\mu)}(g)\,\chi^{(\nu)*}(g) = \delta^{\mu\nu}$$

can be rewritten as

$$\frac{1}{|G|}\sum_{i=1}^{k} p_i\,\chi_i^{(\mu)}\,\chi_i^{(\nu)*} = \delta^{\mu\nu}.$$

Here the summation over group elements $g \in G$ has been replaced by a summation over conjugacy classes. This is therefore an orthogonality relation for the vectors $\left(\sqrt{p_i}\,\chi_i^{(\mu)}\right)$, (with no summation convention implied), in a vector space of dimension k. Since one cannot have more than k linearly independent basis vectors in a k-dimensional space, it follows that the number of inequivalent irreducible representations must be less than or equal to the number of conjugacy classes.

Let

$$r = \text{number of irreducible representations.}$$

hence the previous statement is equivalent to the following inequality

$$r \le k,$$

where k is the number of conjugacy classes.

One can furthermore prove that [3]

$$\frac{1}{|G|}\sum_i p_i\,\chi_i^{(\mu)}\,\chi_i^{(\nu)*} = \delta_{\mu\nu}.$$

Here the index i labels the conjugacy classes, and p_i is the number of elements in the i-th conjugacy class.

This shows that the characters form vectors in a k-dimensional space:

$$\{\chi_1^{(\mu)}, \chi_2^{(\mu)}, \ldots, \chi_k^{(\mu)}\},$$

and that these vectors are mutually orthogonal.

Therefore, the dimension of the space spanned by the conjugacy classes must be less than or equal to the dimension of the space spanned by the irreducible representations:

$$k \le r,$$

where r is the number of irreducible representations, which now sets the dimension of the vector space.

Combining this with the previously obtained inequality $k \geq r$, we conclude that

$$k = r.$$

Hence, the number of inequivalent irreducible representations of a finite group is equal to the number of conjugacy classes of the group. In other words, conjugacy classes and irreducible representations are in one-to-one correspondence.

11.4 Character Decomposition

The orthogonality of characters provides a systematic method to decompose any representation into irreducible components.

For a finite or compact group, every reducible representation is *completely reducible* into a direct sum of irreducible representations. That is, the representation matrices can be brought into a block-diagonal form, where the non-zero blocks coincide with the irreducible representations. This decomposition into block-diagonal form is tightly connected with the decomposition of a the character associated with the reducible representations as a linear combination of the characters of the irreps. The latter, being mutually orthogonal as we have shown above, form the bases of a vector space.

A given irreducible representation may appear more than once in this decomposition. For example, a representation of dimension 5 may decompose into the trivial representation (1) and into an irreducible representation of dimension 2 repeated twice.

In general,

$$D = \sum_{\nu} a_{\nu} \, D^{(\nu)},$$

where a_{ν} is an integer that indicates how many times the irreducible representation $D^{(\nu)}$ appears. In physical applications, these multiplicities are often important.

To determine the coefficients a_{ν}, one takes the trace of the previous equation. This gives

$$\chi(g) = \sum_{\nu} a_{\nu} \, \chi^{(\nu)}(g).$$

Given the block-diagonal structure, the direct sum becomes a normal sum.
Taking the trace, we multiply by $\chi^{(\mu)}(g^{-1})$ and sum over g:

$$\sum_{g} \chi^{(\mu)}(g^{-1}) \, \chi(g) = \sum_{\nu} a_{\nu} \sum_{g} \chi^{(\mu)}(g^{-1}) \, \chi^{(\nu)}(g).$$

Using the orthogonality relations of characters, this gives

$$\sum_{\nu} a_{\nu}\,|G|\,\delta^{\mu\nu} = a_{\mu}\,|G|.$$

Therefore,

$$a_{\mu} = \frac{1}{|G|}\sum_{g} \chi(g)\,\chi^{(\mu)}(g^{-1}),$$

which is important for calculating the decomposition of a reducible representation into irreducible ones:

$$D = \bigoplus_{\mu} a_{\mu}\,D^{(\mu)}.$$

In analogy with standard vectors in vector spaces, e.g.

$$u_i = (\mathbf{e}_i, \mathbf{u}), \qquad \mathbf{u} = u_i\mathbf{e}_i.$$

one can write the expansion of the character as

$$\chi = \sum_{\mu} a_{\mu}\,\chi^{(\mu)},$$

which is the expansion of the vector χ in the basis $\chi^{(\mu)}$.

The coefficients of the projection onto the basis vectors are thus given by

$$a_{\mu} = \langle \chi^{(\mu)}, \chi \rangle,$$

which is directly analogous to projecting a vector onto an orthonormal basis, in the theory of vector spaces.

11.5 The Regular Representation

A particularly important example, which encodes the full structure of the group, is the regular representation.

Starting from cyclic groups C_n and dihedral groups D_n, one can construct them as subgroups of the group of permutations of the n vertices of a polygon.

Indeed, Cayley's theorem states that every finite group of order n is isomorphic to a subgroup of the symmetric group S_n, the latter denoting the permutation group of order n. From the representation-theoretic viewpoint, this leads naturally to a permutation representation.

This can be seen from the fact that, for every finite group G, the multiplication of all its elements $\{a_i\}$ by a fixed element $g \in G$ simply permutes them:

$$\{ga_1, ga_2, \ldots, ga_n\} = \{a_{\pi_1}, a_{\pi_2}, \ldots, a_{\pi_n}\}.$$

Thus, to each element g of the group there corresponds a permutation $\pi(g)$ in S_n:

$$g \longmapsto \pi(g) = \begin{pmatrix} 1 & 2 & \cdots & n \\ \pi_1 & \pi_2 & \cdots & \pi_n \end{pmatrix}.$$

Moreover, the group product is preserved:

$$g_1 g_2 \longleftrightarrow \pi(g_1)\pi(g_2),$$

so the mapping preserves the group structure.

Translating into the language of representations, one has

$$g\, g_i = \sum_j D_{ji}(g)\, g_j,$$

(where g_i is understood as a basis vector).

The permutation matrices $D_{ji}(g)$, of dimension $|G| \times |G|$, form a $|G|$-dimensional representation of the group, called the *regular representation*.

For example, $g\, g_i$ gives a single group element $g_k \in G$, so that

$$g\, g_i = g_k,$$

except for the zero element. Therefore, the matrix $D_{ji}(g)$ has only one non-zero element in each row and in each column.

In the case $g \neq e$, the non-zero elements lie off the diagonal, whereas for $g = e$ they lie on the diagonal; indeed,

$$D_{ji}(e) = \delta_{ji}.$$

11.5.1 Example: Regular Representation for C_3

The cyclic group $C_3 = \{e, c, c^2\}$, which is the symmetry group of equilateral triangles in 2D, was introduced in Chap. 7. It satisfies

$$c^3 = e.$$

It has three elements: the identity e, the generator c (an anti-clockwise rotation by $2\pi/3$), and its square c^2.

In the regular representation, each group element acts by *left multiplication* on the basis of group elements themselves.

We choose the basis vectors:

$$|e\rangle,\ |c\rangle,\ |c^2\rangle.$$

Each is a distinct element of the group written as a vector label.

By definition, the regular representation is given by

$$D(g)\,|h\rangle = |gh\rangle.$$

So for $g = c$ we have:

$$D(c)|e\rangle = |c\rangle,$$

$$D(c)|c\rangle = |c^2\rangle,$$

$$D(c)|c^2\rangle = |e\rangle.$$

Intuitively, the element c acts as to permute the vertices of an equilateral triangle in such as way that this corresponds to an anti-clockwise rotation by $2\pi/3$, see Fig. 11.1.

We now express these results in the chosen basis $(|e\rangle, |c\rangle, |c^2\rangle)$.

Each column of the matrix $D(c)$ tells us where one basis vector is sent under the action of $D(c)$:

Column	Action on	Resulting vector	Coordinates in $(\|e\rangle, \|c\rangle, \|c^2\rangle)$
1	$D(c)\|e\rangle$	$\|c\rangle$	$(0, 1, 0)^T$
2	$D(c)\|c\rangle$	$\|c^2\rangle$	$(0, 0, 1)^T$
3	$D(c)\|c^2\rangle$	$\|e\rangle$	$(1, 0, 0)^T$

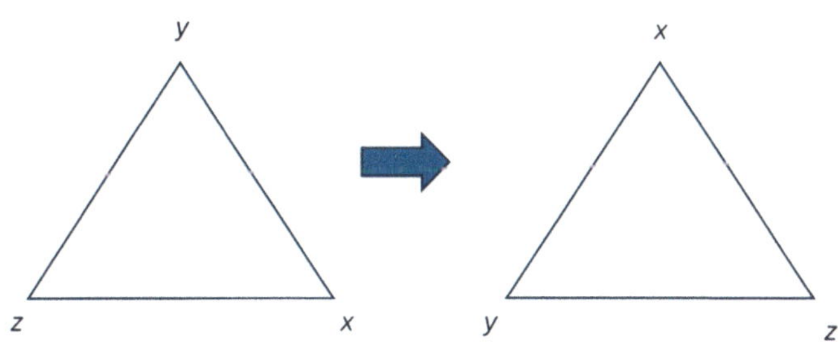

Fig. 11.1 Action of the group element ($c \in C_3$) viewed as an element of the permutation group, acting by cyclic permutation of the vertices of an equilateral triangle

We therefore build the matrix by placing these columns side by side:

$$D(c) = \begin{pmatrix} 0\,0\,1 \\ 1\,0\,0 \\ 0\,1\,0 \end{pmatrix}.$$

The matrix $D(c)$ cyclically permutes the basis vectors:

$$|e\rangle \longrightarrow |c\rangle \longrightarrow |c^2\rangle \longrightarrow |e\rangle.$$

Thus, $D(c)$ is a cyclic permutation matrix.

This is exactly what the regular representation of the generator of C_3 looks like. Above we used a basic principle of linear operators in linear algebra: Because of how matrix-vector multiplication works, for a linear operator T and a generic vector written in the (generic) basis $(\mathbf{e}_1, \mathbf{e}_2)$,

$$\left(T\right)\begin{pmatrix} a \\ b \end{pmatrix} = a\,T(\mathbf{e}_1) + b\,T(\mathbf{e}_2).$$

Therefore, the matrix representing T must contain the images of the basis vectors as its columns. This is what ensures that linearity works correctly: each column of the matrix corresponds to the action of T on one basis vector.

11.5.2 Decomposition of the Regular Representation into Irreps

Now let us decompose the regular representation into its irreducible components:

$$D = \sum_{\nu} a_{\nu}\, D^{(\nu)}.$$

The coefficients are given by

$$a_{\mu} = \frac{1}{|G|} \sum_{g} \chi(g)\, \chi^{(\mu)}(g^{-1}),$$

which can be computed using the orthogonality relations of characters.

From the explicit form of the matrices $D_{ji}(g)$, we deduce that the character of the regular representation is

$$\chi(g) = \begin{cases} 0, & g \neq e, \\ |G|, & g = e. \end{cases}$$

This is the character of the regular representation. Its simplicity makes the regular representation a powerful bookkeeping tool.

The character of the identity element $\chi(e)$ is always equal (as for any representation) to the dimension of the representation. In the case of the regular representation, the dimension is $|G|$.

Therefore,

$$a_\mu = \frac{1}{|G|}\,\chi(e)\,\chi^{(\mu)}(e) = \chi^{(\mu)}(e) = n_\mu,$$

where n_μ is the dimension of the irreducible representation $D^{(\mu)}$.

Thus, each irreducible representation appears in the regular representation a number of times equal to its dimension. This is a fundamental fact that we are going to use to deduce another basic tool in the making of character tables. Here n_μ is the dimension of the irreducible representation $D^{(\mu)}$.

Therefore, setting $g = e$ in

$$\chi(g) = \sum_\mu n_\mu\,\chi^{(\mu)}(g),$$

we obtain

$$|G| = \sum_\mu n_\mu\,\chi^{(\mu)}(e) = \sum_\mu n_\mu^2.$$

Hence,

$$|G| = \sum_\mu n_\mu^2 \tag{11.2}$$

This identity is one of the most important consistency checks in character theory.

On the other hand, for all $g \neq e$ we obtain

$$0 = \sum_\mu n_\mu\,\chi^{(\mu)}(g). \tag{11.3}$$

Combining Eqs. (11.2) and (11.3), we finally obtain

$$\frac{1}{|G|}\sum_\mu \chi^{(\mu)}(e)\,\chi^{(\mu)}(g) = \begin{cases} 0, & g \neq e, \\ 1, & g = e, \end{cases}$$

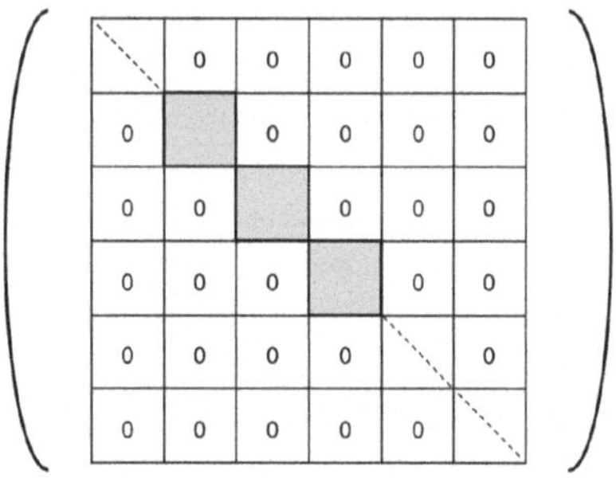

Fig. 11.2 Schematic block-diagonal structure of the regular representation after decomposition into irreducible components. Each irreducible representation $D^{(\mu)}$ of dimension n_μ appears n_μ times along the diagonal, while all other matrix elements vanish

where we have used

$$n_\mu = \chi^{(\mu)}(e).$$

Therefore, for the *regular representation*:

The number of times that an irreducible representation of a group appears inside the regular representation is equal to the dimension of the irrep.

An irrep $V^{(\mu)}$ of dimension n_μ appears n_μ times inside the regular representation matrix, and it is described by a matrix $D^{(\mu)}(g)$ of dimension $n_\mu \times n_\mu$.

The sketch in Fig. 11.2 illustrates the block-diagonal structure: in the block-diagonal decomposition of the regular representation into irreps, the same $n_\mu \times n_\mu$ block is repeated n_μ times along the diagonal. Each block corresponds to an independent irreducible sector.

Example: Regular Representation Decomposition for C_3
Going back to the previous example of the group C_3, our matrix

$$D(c) = \begin{pmatrix} 0 & 0 & 1 \\ 1 & 0 & 0 \\ 0 & 1 & 0 \end{pmatrix}$$

is a 3×3 permutation matrix: the regular representation of the generator c. Hence, the regular representation of C_3 is of dimension three, which is larger than the dimension of the irreps (they are all of dimension one because the group is abelian).

If we diagonalize this matrix (e.g. via Fourier transform on C_3), we obtain:

$$D(c)_{\text{diag}} = \begin{pmatrix} 1 & 0 & 0 \\ 0 & \omega & 0 \\ 0 & 0 & \omega^2 \end{pmatrix}, \qquad \omega = e^{2\pi i/3}.$$

This shows explicitly that the regular representation decomposes as

$$R \simeq \mathbf{1} \oplus \mathbf{1}' \oplus \mathbf{1}''.$$

Here the symbols $\mathbf{1}, \mathbf{1}', \mathbf{1}''$ denote the three inequivalent one-dimensional irreducible representations of the group C_3. The bold symbol $\mathbf{1}$ indicates a representation of dimension one, while the primes distinguish different irreducible representations of the same dimension. Explicitly, they correspond to the characters

$$\mathbf{1}: \quad D(c) = 1, \qquad \mathbf{1}': \quad D(c) = \omega, \qquad \mathbf{1}'': \quad D(c) = \omega^2,$$

with $\omega = e^{2\pi i/3}$.

Three one-dimensional blocks correspond to exactly one copy of each irreducible representation. This decomposition is directly visible as diagonal blocks, where the blocks are just numbers in $\mathbb{C}$ because the group is abelian.

The regular representation thus serves as a unifying framework that contains every irreducible representation with multiplicity equal to its dimension, providing a complete and elegant summary of the group's representation theory.

References

1. A. Zee, *Group Theory in a Nutshell for Physicists* (Princeton University Press, Princeton, 2016)
2. J.F. Cornwell, *Group Theory in Physics*, vol. 1 (Academic Press, London, 1997)
3. M. Hamermesh, *Group theory and its application to physical problems*. Addison-Wesley Series in Physics (Addison-Wesley Publishing Company, Reading, 1962)

Chapter 12
Character Tables

Abstract In this chapter we will learn how to use the tools developed in the previous chapter in order to build character tables for finite groups. In particular, we will provide a set of rules to compute characters in order to fill in the character table of a given group. We shall build the character tables of group C_3 (abelian) and D_3 (non-abelian) from scratch and step by step. We will also begin to learn how to use the character table to implement the decomposition of a reducible representation into irreps. An excellent reference for this part is, again, Jones [1]. This chapter translates the abstract orthogonality results of representation theory into a concrete algorithmic procedure, allowing character tables to be constructed explicitly and used in practical symmetry analysis.

12.1 Rules for the Construction of the Character Table

In the previous chapter we established the orthogonality relations for irreducible characters and showed how they encode the structure of a group. We now show how these results can be turned into a systematic procedure for constructing character tables.

The *character table* of a finite group corresponds to a table in which the rows represent its irreducible representations and the columns represent its conjugacy classes; its entries therefore correspond to the class characters $\chi_\mu(c)$. Each row captures how a specific irreducible representation "sees" the group, while each column groups together symmetry operations that are equivalent under conjugation.

In order to construct the character table of a group, it is useful to keep in mind some fundamental properties derived from the Great Orthogonality Theorem, which have already been proved:

In the character table:

- the rows correspond to the irreducible representations (irreps),
- the columns correspond to the conjugacy classes of the group.

A. Zaccone, *Differential Geometry and Group Theory*,
https://doi.org/10.1007/978-3-032-27518-9_12

Taken together, these properties form a closed set of constraints that uniquely determine the character table.

The tools used to construct the character table are the following:

1. The number of irreducible representations equals the number of conjugacy classes:
$$r = k.$$
2. The sum of the squares of the dimensions of the irreducible representations equals the order of the group:
$$\sum_{\mu} m_{\mu}^{2} = |G|.$$
3. Orthogonality of characters:
$$\sum_{i} p_i \, \chi_i^{(\mu)} \, \chi_i^{(\nu)*} = |G| \, \delta_{\mu\nu},$$
where p_i is the number of elements in the i-th conjugacy class.
4. Any additional available information. For example, characters obey the group multiplication properties. For one-dimensional irreducible representations the representation matrix is simply a complex number. Therefore the character, being the trace of the representation matrix, coincides with the representation itself: $\chi(g) = D(g)$.

Example: Abelian Groups
Abelian groups provide the simplest testing ground for the character table machinery, since all group elements commute.

All irreducible representations of an abelian group are one-dimensional. Indeed, if G is abelian then
$$D(g)D(h) = D(h)D(g) \qquad \forall g, h \in G.$$
Thus every matrix of the representation commutes with all the others. By Schur's lemma, in an irreducible representation any operator commuting with the representation must be proportional to the identity. Therefore the representation space cannot have dimension larger than one, and all irreducible representations are one-dimensional.

As a consequence, all matrices $D^{(\mu)}(g)$ are diagonal. Therefore, if the dimension of the representation is greater than one, the representation is reducible, and one eventually finds only one-dimensional irreducible representations. Equivalently, all irreducible representations of an Abelian group are characters.

More generally, the representation matrices can be written in block-diagonal form, with diagonal blocks. This reflects the decomposition of the vector space into invariant one-dimensional subspaces.

12.2 The Character Table of Group C_3

We now apply the general construction rules to the simplest nontrivial finite group.

Consider the group $C_3 = \{e, c, c^2\}$, where

$$c = 120^\circ, \qquad c^2 = 240^\circ.$$

In this example, we have three conjugacy classes,

$$\{e\}, \quad \{c\}, \quad \{c^2\},$$

and therefore we expect three irreducible representations. This immediately fixes the size of the character table.

We should recall from Chap. 7 that, for abelian groups, each element forms its own conjugacy class. Indeed, for any $g, a \in G$,

$$gag^{-1} = a, \ \forall g \in G$$

since all elements commute. Indeed, the character table must *always* be a square table (as long as only irreps are involved). As we know already that there are three conjugacy classes, hence also the irreps must be three. We denote the irreducible representations of C_3 by: $D^{(1)}$, $D^{(2)}$, $D^{(3)}$. Thus, the character table has three rows and three columns.

The group is abelian; therefore all irreducible representations are one-dimensional and must satisfy the group multiplication law. In particular,

$$\chi(c^2) = [\chi(c)]^2, \qquad \chi(c)^3 = \chi(c^3) = \chi(e) = 1.$$

Therefore, we deduce from the last line that the characters must be the cubic roots of unity. This reflects the fact that one-dimensional representations of cyclic groups are homomorphisms into the complex unit circle. Using Euler's formula,

$$\omega = e^{2\pi i/3}, \qquad \omega^2 = e^{4\pi i/3}.$$

The representation $D^{(1)}$ is the trivial representation, in which every group element is mapped to the identity. Hence, now that we know all the characters, we can now populate the character table, for example in this way:

C_3	e	c	c^2
$D^{(1)}$	1	1	1
$D^{(2)}$	1	ω	ω^2
$D^{(3)}$	1	ω^2	ω

This table fully encodes how all irreducible representations of C_3 respond to the group's symmetry operations.

In crystallographic notation one writes $A \equiv D^{(1)}$; the representations $D^{(2)}$ and $D^{(3)}$ are complex conjugates of each other.

A trivial irreducible representation is the one in which a scalar is not affected in any way by the action of the group. For example, the z-component of a three-dimensional vector.

In many physical applications, one is interested in how reducible representations decompose into irreducible ones.

Of course, one may choose to include also reducible representations, in which case the character table is no longer square. For example, we may want to include the reducible vector representation of C_3 which is irreducible in $\mathbb{R}$ but of course reducible in $\mathbb{C}$. In this case we have the following table:

C_3	e	c	c^2
A	1	1	1
E	2	$2\cos\left(\frac{2\pi}{3}\right)$	$2\cos\left(\frac{4\pi}{3}\right)$
Γ^+	1	$e^{i2\pi/3}$	$e^{i4\pi/3}$
Γ^-	1	$e^{-i2\pi/3}$	$e^{-i4\pi/3}$
V	3	0	0

The representation A is the trivial representation and corresponds to a scalar, for example the z-component of a three-dimensional vector, or the combination x^2+y^2 in $\mathbb{R}^2$.

The representation E is two-dimensional and acts in the xy-plane, with character

$$\chi^E(c) = 2\cos\varphi.$$

It describes planar rotations, for instance through the combinations $x \pm iy$.

The representations Γ^+ and Γ^- are one-dimensional irreducible representations of the abstract group C_3 over the complex field $\mathbb{C}$, defined by the cubic roots of unity.

The three-dimensional, hence reducible, vector representation V decomposes as

$$V = A \oplus E.$$

While $\Gamma^\pm$ require complex scalars, the representation E can be regarded as an irreducible representation over the real vector space $\mathbb{R}^2$.

This decomposition can be computed as follows. We have already seen that the vector representation of a three-dimensional vector (x, y, z) is given by the matrices

$$D(e),$$

$$D(c) = R\left(\tfrac{2\pi}{3}\right) = \begin{pmatrix} -\frac{1}{2} & -\frac{\sqrt{3}}{2} & 0 \\ \frac{\sqrt{3}}{2} & -\frac{1}{2} & 0 \\ 0 & 0 & 1 \end{pmatrix},$$

$$D(c^2) = R\left(\tfrac{4\pi}{3}\right) = \begin{pmatrix} -\frac{1}{2} & \frac{\sqrt{3}}{2} & 0 \\ -\frac{\sqrt{3}}{2} & -\frac{1}{2} & 0 \\ 0 & 0 & 1 \end{pmatrix}.$$

This representation can be decomposed into irreducible representations using the characters relations. The character for the reducible vector representation is

$$\chi^V = \left\{\chi^V(e), \chi^V(c), \chi^V(c^2)\right\}$$

which is the "composite" character. As such it can be written as a linear combination of the "basis vectors" (the characters play the role of basis vectors in a finite-dimensional inner-product space), i.e. the characters of the irreps:

$$\chi^V(g) = \sum_\nu a_\nu \, \chi^{(\nu)}(g),$$

or

$$\chi^V = a_1 \, \chi^{(1)} + a_2 \, \chi^{(2)} + a_3 \, \chi^{(3)},$$

with coefficients given by the orthogonality theorem. The latter gives a system of simultaneous equations for the coefficients a_ν. In fact, the set of independent equations is equal to the number of conjugacy classes, since elements belonging to the same conjugacy class have the same value of the character χ.

The coefficients are given by

$$a_\nu = \frac{1}{|G|} \sum_i p_i \, [\chi_i^{(\nu)}]^* \, \chi_i,$$

where the sum runs over conjugacy classes.

Thus,

$$a_\nu = \frac{1}{|G|} \sum_i p_i \, [\chi_i^{(\nu)}]^* \, \chi_i^V.$$

In the case of C_3,

$$\chi^V = \{\chi^V(e), \chi^V(c), \chi^V(c^2)\} = (3, 0, 0).$$

Hence, for each irrep $\nu = 1, 2, 3$, reading from left to right along the row, we compute

$$a^{(1)} = \frac{1}{3}\big[1 \cdot 1 \cdot 3 + 1 \cdot 1 \cdot 0 + 1 \cdot 1 \cdot 0\big] = \frac{1}{3}\big(3\,\chi^{(1)}(e)\big),$$

$$a^{(2)} = \frac{1}{3}\big[1 \cdot 1 \cdot 3 + 1 \cdot \omega \cdot 0 + 1 \cdot \omega^2 \cdot 0\big] = \frac{1}{3}\big(3\,\chi^{(2)}(e)\big),$$

$$a^{(3)} = \frac{1}{3}\big[1 \cdot 1 \cdot 3 + 1 \cdot \omega^2 \cdot 0 + 1 \cdot \omega \cdot 0\big] = \frac{1}{3}\big(3\,\chi^{(3)}(e)\big).$$

Because $\chi^{(3)}(e) = 1$, all the coefficients are equal to one, and therefore

$$a_\nu = \langle \chi^V, \chi^{(\nu)} \rangle = \frac{1}{3}\,\big(3\,\chi^{(\nu)}(e)\big) = 1.$$

$$\Rightarrow \quad \chi^V = \chi^{(1)} + \chi^{(2)} + \chi^{(3)},$$

and we obtain the decomposition of the vector representation into a direct sum:

$$D^V = D^{(1)} \oplus D^{(2)} \oplus D^{(3)} = \Gamma_0 \,\oplus\, \Gamma^+ \,\oplus\, \Gamma^-.$$

where we switched to the solid-state physics notation after the last equality. Here

$$\Gamma^+ \,\oplus\, \Gamma^- = E$$

forms the two-dimensional irreducible representation E (in crystallographic notation), while Γ_0 is the trivial representation (A) (in crystallographic notation). Hence, we have, in crystallographic notation, simply:

$$D^V = A \,\oplus\, E.$$

Having mastered the abelian case, we now turn to a genuinely non-abelian group, where higher-dimensional irreducible representations necessarily appear.

12.3 The Character Table of Group D_3

The group D_3 provides the simplest example in which non-commuting symmetry operations fundamentally alter the structure of the character table.

We shall now consider a finite group which is non-abelian. The most logic choice, after we have mastered the case of group C_3 is to turn our attention to group D_3 where the out-of-plane (dihedral) rotations of π do not commute (as is the case for rotations in 3D).

The group is formed by in-plane rotations analogous to those of C_3 plus out-of-plane dihedral rotations by 180° about the axes AX, BY, CZ, cf. Fig. 12.1. The latter are equivalent to reflections in C_{3v} through planes vertical to the page and passing through AX, BY, CZ.

We denote the rotations by 180° about AX, BY, CZ by b_1, b_2, b_3: they are reflections with respect to the three medians.

They are *two-fold rotations*, i.e. they satisfy

$$b_i^2 = e,$$

and the corresponding axes are called two-fold symmetry axes. Note that: $c(AO) = BO$. Also, a rotation by $2\pi/3$ maps AO into BO. This implies that b_2 and b_1 are related by conjugation with c:

$$b_2 = c\, b_1\, c^{-1}.$$

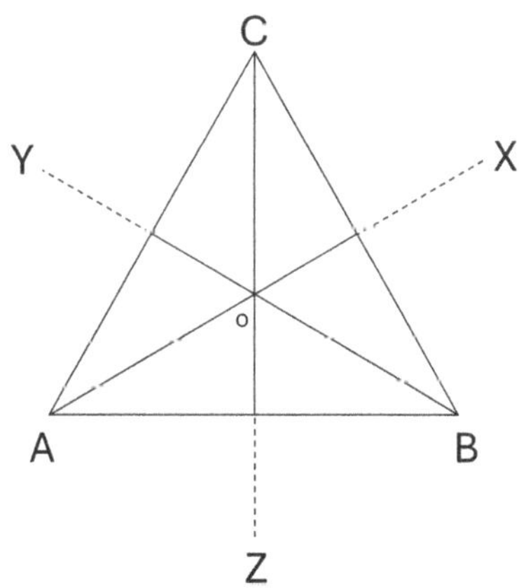

Fig. 12.1 Geometric representation of the symmetry operations of the dihedral group D_3. The equilateral triangle illustrates the rotational symmetries about the center o: the identity and the rotations by (120°) and (240°) around the vertical axis, are just the same as e, c and c^2 in group C_3. The bisector lines indicate the three reflection axes passing through o, which correspond to the three out-of-plane rotations by 180° (not included in C_3). Together, these rotations and reflections generate the full symmetry group D_3 of the triangle

The effect of b_2 can be reproduced by first applying the inverse of c, then performing a rotation by π, and finally applying again a rotation by $2\pi/3$.

Note: we consider the axis OY fixed in space.

The conjugacy (equivalence) classes are

$$(e), \qquad (c, c^2), \qquad (b_1, b_2, b_3),$$

which we denote by K_1, K_2, K_3. The geometric interpretation of the symmetry operations, cf. Sect. 7.4.1 in Chap. 7, makes the conjugacy structure particularly transparent.

There are therefore three conjugacy classes, hence three irreducible representations. Moreover,

$$\sum_{\mu} n_{\mu}^2 = |G| \quad \Rightarrow \quad n_1^2 + n_2^2 + n_3^2 = 6.$$

But the trivial representation is

$$\Gamma_0 \equiv D^{(1)} = 1, \qquad n_1 = 1.$$

Therefore

$$n_2^2 + n_3^2 = 5.$$

This severely restricts the possible dimensions of the remaining irreducible representations.

The only solution is

$$n_2 = 1, \qquad n_3 = 2 \quad \text{(or vice versa)}.$$

This allows us to fill in the first column of the character table, since

$$\chi^{(\mu)}(e) = n_{\mu}.$$

For a one-dimensional representation the representation matrices reduce to complex numbers, so that the homomorphism property gives

$$D(bc) = D(b)D(c).$$

Taking the trace (which in one dimension coincides with the matrix itself) one obtains

$$\chi(bc) = \chi(b)\chi(c).$$

But since b and bc belong to the same conjugacy class, we also have

$$\chi(b) = \chi(bc),$$

because elements belonging to the same conjugacy class have the same character.

Therefore

$$\chi(c) = \chi_2 = 1.$$

Note: this value of $\chi(c)$ refers to the one-dimensional representation of the in-plane rotation c.

We then find

$$\chi(b) = \chi(b^2) = \chi(e) = 1,$$

which gives $\chi_3 = \pm 1$.

The +1 solution corresponds to the trivial representation, $D^{(1)}$. Hence, for $D^{(2)}$ we must have the other solution,

$$\chi_3 = -1.$$

Thus we have used properties (i), (ii), and (iv) to determine some of the characters for D_3. In particular, we have been able to fill in the first two rows of the character table: Thus we have used properties (i), (ii), and (iv) to determine the character table:

D_3	K_1	K_2	K_3
$D^{(1)}$	1	1	1
$D^{(2)}$	1	1	-1
$D^{(3)}$	2	α	β

In crystallographic notation, the irreducible representations are denoted by

$$A_1, \quad A_2, \quad E,$$

where E is two-dimensional.

Typical basis functions are:

$$A_1 : x^2 + y^2,\ x^2 + y^2 + z^2, \qquad A_2 : z, \qquad E : (x, y).$$

The remaining unknown entries are fixed uniquely by orthogonality.

Indeed, to determine the remaining characters of $D^{(3)}$, for which we used placeholder variables α and β in the above table, we shall now use the orthogonality relations:

$$\frac{1}{|G|} \sum_i p_i \, \chi^{(\mu)}(K_i) \, \chi^{(\nu)}(K_i)^* = \delta^{\mu\nu}.$$

First, we consider the scalar product between the first and the third rows:

$$\langle \chi_1, \chi_3 \rangle = \frac{1}{6}\big(1 \cdot 1 \cdot 2 + 2 \cdot 1 \cdot \alpha + 3 \cdot 1 \cdot \beta\big) = 0.$$

Similarly, for the scalar product between the second and third rows, we have:

$$\langle \chi_2, \chi_3 \rangle = \frac{1}{6}\big(1 \cdot 1 \cdot 2 + 2 \cdot 1 \cdot \alpha + 3 \cdot (-1) \cdot \beta\big) = 0.$$

Thus we obtain

$$\begin{cases} 2 + 2\alpha + 3\beta = 0, \\ 2 + 2\alpha - 3\beta = 0, \end{cases} \qquad \Rightarrow \qquad \begin{cases} \beta = 0, \\ \alpha = -1. \end{cases}$$

Thus, the character table is completely determined.
Note: Conjugate rotations are rotations by the same angle, but the converse is not always true. The class K_2 contains two elements, both of which are rotations by $120°$, which generate the group C_3. For this reason, K_2 coincides with C_3. The class K_3, on the other hand, contains three two-fold out-of-plane rotations. Hence it consists of the three elements of type C_2.

Alternatively, we could have determined more of the table characters using the orthogonality relations. For example, we could have written the table with placeholder variables as follows:

D_3	(e)	(c, c^2)	(b, bc, bc^2)
$D^{(1)}$	1	1	1
$D^{(2)}$	1	α	β
$D^{(3)}$	2	γ	δ

Using the orthogonality relations for rows:

$$\langle \chi_1, \chi_2 \rangle \;\Rightarrow\; \frac{1}{6}\big(1 + 2\alpha + 3\beta\big) = 0,$$

$$\langle \chi_2, \chi_2 \rangle \;\Rightarrow\; \frac{1}{6}\big(1 + 2\alpha^2 + 3\beta^2\big) = 1.$$

Using the orthogonality relations for columns:

$$\frac{1}{|G|} \sum_{\nu} \chi^{(\nu)}(C_i)\, \chi^{(\nu)}(C_j)^* = \delta_{ij},$$

which yields

$$\frac{1}{6}(1+\alpha+2\gamma)=0, \qquad \frac{1}{6}(1+\beta+2\delta)=0.$$

In total, we have four equations in four unknowns. However, we can also use the group multiplication rules to select the physically consistent solution. One finds

$$\alpha^2=1, \qquad \beta=1.$$

Since χ_2 is a 1-dimensional character, $|\alpha|=|\beta|=1$ and in fact $\alpha^2=\beta^2=1$ here. Hence $\alpha,\beta\in\{\pm1\}$. Plugging into $1+2\alpha+3\beta=0$ gives the unique solution:

$$\alpha=1, \qquad \beta=-1.$$

Column orthogonality (sum over irreps) gives:

$$\frac{1}{6}\big(1+\alpha+2\gamma\big)=0 \;\Rightarrow\; 1+\alpha+2\gamma=0 \;\Rightarrow\; \gamma=-1,$$

$$\frac{1}{6}\big(1+\beta+2\delta\big)=0 \;\Rightarrow\; 1+\beta+2\delta=0 \;\Rightarrow\; \delta=0.$$

And the final character table follows as:

D_3	(e)	(c,c^2)	(b,bc,bc^2)
$D^{(1)}\,(A_1)$	1	1	1
$D^{(2)}\,(A_2)$	1	1	-1
$D^{(3)}\,(E)$	2	-1	0

This table summarizes all irreducible symmetry behaviors of the equilateral triangle.

Once the character table is known, decomposition of any representation becomes a straightforward calculation.

Also in this case, we can use the character table to determine how the vector representation acting on the basis (x,y,z) decomposes into the various irreducible representations.

To this end, we need the matrices of one of the group elements. Without loss of generality, we may choose a twofold rotation about the x axis.

The corresponding matrix representation is

$$D^{(V)}(b)=\mathrm{diag}(1,-1,-1),$$

which implements the transformation

$$x'=x, \qquad y'=-y, \qquad z'=-z.$$

Consequently, the character of this element in the vector representation is

$$\chi^{(V)}(b) = -1.$$

Therefore, we obtain the character vector

$$\chi^{(V)} = (3,\ 0,\ -1),$$

corresponding to the characters $\chi(e)$, $\chi(c)$, and $\chi(b)$.

To find the coefficients of the decomposition, we proceed as usual using the character table.

To this end, we need the matrices of one of the group elements. Without loss of generality, we may choose a twofold rotation about the x axis.

The corresponding matrix representation is

$$D^{(V)}(b) = \mathrm{diag}(1, -1, -1),$$

which implements the transformation

$$x' = x, \qquad y' = -y, \qquad z' = -z.$$

Consequently, the character of this element in the vector representation is

$$\chi^{(V)}(b) = -1.$$

Therefore, we obtain the character vector

$$\chi^{(V)} = (3,\ 0,\ -1),$$

corresponding to the characters $\chi(e)$, $\chi(c)$, and $\chi(b)$.

To find the coefficients of the decomposition, we proceed as usual using the character table. To find the coefficients of the decomposition, as usual we use

$$a_\mu = \frac{1}{|G|} \sum_i p_i\, \chi^{(\nu)}(g_i)\, \chi^{(\mu)}(g_i)^*.$$

For the irreducible representation A_1 we obtain

$$a_{A_1} = \frac{1}{6}\big(1 \cdot 1 \cdot 3 + 2 \cdot 1 \cdot 0 + 3 \cdot 1 \cdot (-1)\big) = 0.$$

For the irreducible representation A_2 we obtain

$$a_{A_2} = \frac{1}{6}\big(1 \cdot 1 \cdot 3 + 2 \cdot 1 \cdot 0 + 3 \cdot (-1) \cdot (-1)\big) = 1.$$

For the irreducible representation E we obtain

$$a_E = \frac{1}{6}\big(1 \cdot 2 \cdot 3 + 2 \cdot (-1) \cdot 0 + 3 \cdot 0 \cdot (-1)\big) = 1.$$

Therefore,

$$\chi^{(V)} = \chi^{(A_2)} + \chi^{(E)},$$

and the vector representation decomposes as

$$D^{(V)} = A_2 \oplus E.$$

The z component of the three-dimensional vector (x, y, z) changes sign, picks up a (-1) factor, under the operation b, but remains invariant under c $(+1)$; it therefore forms the natural basis of the irreducible representation A_2.

The two-dimensional representation E describes the transformations of x and y; however, unlike the case of C_3, once the additional rotations b_i are included, it can no longer be reduced further.

This illustrates how character tables directly reveal the symmetry-adapted structure of physical degrees of freedom.

Character tables thus provide a complete and practical summary of the representation theory of a finite group, serving as the main computational tool for symmetry analysis in physics and chemistry.

Exercise

Determine the irreducible representations contained in the representation of the group D_3 acting on the space of the functions

$$x^2,\ y^2,\ xy$$

(note that these can be used as a basis for the representation).

Determine the coefficients a_μ of the decomposition of the representation D into its irreducible components (A_1, A_2, E).

Reference

1. H.F. Jones, *Groups, representations and physics*, 2nd edn. (Taylor & Francis Ltd, Bristol/New York, 1998)

Chapter 13
Tensor Product Representation

Abstract In this chapter we take a closer look at the direct products or tensor products of representations of groups. This situation is of special relevance for physics, in particular it is unavoidable whenever we a system formed by two or more particles. Indeed, in that case we are presented with two possibilities: the two, or more, particles, are subjected to the same spatial transformation or they can transform independently of each other. In the first case, the tensor product of irreducible representations of the same group is highly redundant, which means, formally, that it is massively reducible. In the second case, instead, the tensor product representation is irreducible because it matches the effective dimensionality of the system. Again, an excellent reference for this part is the book by Jones [1]. This chapter therefore marks the transition from single-particle symmetry analysis to the representation theory of composite systems, highlighting how group structure controls reducibility, coupling, and emergent symmetry.

13.1 Direct Products of Representations and Their Decomposition

Having developed the machinery of irreducible representations and character tables, we now turn to the problem of combining representations.

Let us consider a system of two particles, for example two electrons, described by wave functions

$$\psi_a(\mathbf{x}), \qquad \psi_b(\mathbf{x}),$$

which transform according to the irreducible representations

$$D^{(\mu)}, \qquad D^{(\nu)},$$

respectively, of some group G. This situation arises ubiquitously in quantum mechanics whenever composite systems are formed.

A. Zaccone, *Differential Geometry and Group Theory*,
https://doi.org/10.1007/978-3-032-27518-9_13

For atoms in a solid, this group could be a point symmetry group, whereas for isolated atoms it could be the rotation group SO(3).

An element $g \in G$ determines a transformation of the wave functions:

$$\psi'_a(\mathbf{x}) = D^{(\mu)}_{ba}(g)\,\psi_b(\mathbf{x}),$$

$$\psi'_c(\mathbf{x}) = D^{(\nu)}_{dc}(g)\,\phi_d(\mathbf{x}).$$

Therefore, the product of the wave functions

$$\psi_{ac}(\mathbf{x}) = \psi_a(\mathbf{x})\,\phi_c(\mathbf{x})$$

transforms as

$$\psi'_{ac}(\mathbf{x}) = D^{(\mu)}_{ba}(g)\,D^{(\nu)}_{dc}(g)\,\psi_{bd}(\mathbf{x}).$$

We can regard the product of the two matrices as a single matrix whose first index is the ordered pair $(bd) = B$ and whose second index is the ordered pair $(ac) = A$. This is simply a convenient relabeling of indices that allows the tensor product to be treated as an ordinary matrix representation. We define

$$D^{(\mu\times\nu)}_{BA}(g) \equiv D^{(\mu)}_{ba}(g)\,D^{(\nu)}_{dc}(g).$$

Thus the transformation law becomes

$$\psi'_A(\mathbf{x}) = D^{(\mu\times\nu)}_{BA}(g)\,\psi_B(\mathbf{x}).$$

This shows that the product wave function transforms according to the *direct product representation*

$$D^{(\mu)} \otimes D^{(\nu)}.$$

In representation theory this construction is also called the tensor product representation.

At this point, it is crucial to distinguish whether the two particles transform independently or are constrained to transform together. If, as a first approximation, we can neglect the interaction between the two electrons, the system has a symmetry group $G \times G$, i.e. it is invariant under independent transformations g_1 and g_2 acting on the two electrons, and the resulting wave functions are given by the above products, suitably anti-symmetrized by the Fermi statistics.

This enlargement of the symmetry group has immediate consequences for irreducibility.

The resulting wave functions are therefore products of the type discussed above, and the representation matrices factorize as

$$(A \otimes B)_{ij,st} = A_{ij}\, B_{st}.$$

The representation matrix is thus given by:

$$D^{(\mu\times\nu)}_{bd;\,ac}(g_1, g_2) \equiv D^{(\mu)}_{ba}(g_1)\, D^{(\nu)}_{dc}(g_2)$$

Here one should note that:

$$D^{(\mu\times\nu)}(g_1, g_2) = D^{(\mu)}(g_1)\, D^{(\nu)}(g_2)$$

If $D^{(\mu)}$ and $D^{(\nu)}$ are irreducible representations of G, then $D^{(\mu)} \otimes D^{(\nu)}$ is an irreducible representation of the tensor product group $G \times G$ (which follows from the fact that each factor acts on an independent degree of freedom):

$$D^{(\mu)}(g_1)\, D^{(\nu)}(g_2) = \big(D^{(\mu)}(g_1) \otimes D^{(\nu)}(g_1)\big)\big(D^{(\mu)}(g_2) \otimes D^{(\nu)}(g_2)\big)$$

$$= \big(D^{(\mu)}(g_1)D^{(\mu)}(g_2)\big) \otimes \big(D^{(\nu)}(g_1)D^{(\nu)}(g_2)\big) = D^{(\mu\times\nu)}(g_1 g_2)$$

where the following rule of tensor products was used:

$$(A \otimes B)(A' \otimes B') = AA' \otimes BB'$$

see [2], Appendix F, p. 274.

The situation changes qualitatively once the two particles are dynamically coupled. When there is interaction between the two electrons, such that they perform the same rotations the relevant symmetry is reduced to that of the original group G, inasmuch as the same rotation is applied to both electrons:

$$g_1 = g_2 = g$$

In this situation, in general the product representation will be reducible. The reduction of $D^{(\mu)} \otimes D^{(\nu)}$ will be:

$$D^{(\mu)} \otimes D^{(\nu)} = \sum_{\sigma} a_\sigma\, D^{(\sigma)}$$

This decomposition is known as the Clebsch–Gordan series and is of great importance in physics, as we shall see also in the next chapters. It plays a central role in angular momentum coupling and many-body quantum systems.

13.2 Decomposition of Tensor Products into Irreps

We now show how tensor-product representations can be reduced systematically using character theory.

We have seen that $D^{(\mu)} \otimes D^{(\nu)}$ is the product of two irreps. It has dimension $M_\mu M_\nu$, which can often be larger than the maximum order of any irreducible representation of G. This rapid growth in dimension makes reducibility unavoidable when only a single group acts.

Therefore, in the case of two interacting or coupled electrons, they are subjected to the same rotation

$$g_1 = g_2 = g,$$

that is, the group is G and the product group will have a representation whose dimensionality exceeds the maximum dimension of an irreducible representation of G. The latter maximum dimension of any irrep is set by the requirement that $\sum_\mu n_\mu^2 = |G|$. In this case, the product must be reducible:

$$D^{(\mu)} \otimes D^{(\nu)} = \sum_\sigma a_\sigma \, D^{(\sigma)} \qquad \text{(Clebsch–Gordan series)}$$

The strategy mirrors exactly the decomposition of reducible representations discussed in the previous chapter.

How are the coefficients a_σ determined? As usual, by means of the same character algebra that we have developed up to now.

The character of $D^{(\mu\times\nu)}(g)$ is:

$$\chi^{(\mu\times\nu)}(g) = D^{(\mu\times\nu)}_{AA}(g) = D^{(\mu)}_{bb}(g)\, D^{(\nu)}_{cc}(g) = \chi^{(\mu)}(g)\, \chi^{(\nu)}(g)$$

where we have set $A = ac$ in

$$D^{(\mu\times\nu)}_{bd;\, ac}(g)$$

and summed over repeated indices:

$$\sum_{ac} D^{(\mu\times\nu)}(g)_{ac,\, ac} = \sum_{ac} D^{(\mu)}_{aa}(g)\, D^{(\nu)}_{cc}(g) = \left(\sum_a D^{(\mu)}_{aa}(g)\right)\left(\sum_c D^{(\nu)}_{cc}(g)\right)$$

$$= \chi^{(\mu)}(g)\, \chi^{(\nu)}(g).$$

Thus, characters of tensor-product representations factorize into products of characters.

In the previous chapter we saw how the composite representation (V) associated with the transformation of a cartesian vector (x, y, z) decomposes into irreps, using

$$a_\nu = \langle \chi^{(\nu)}, \chi^{(V)} \rangle.$$

Here we do the same thing for the tensor-product representation as the composite representation which is being decomposed:

$$a_\sigma = \langle \chi^{(\sigma)}, \chi^{(\mu)} \chi^{(\nu)} \rangle$$

Here $\chi^{(\mu)}\chi^{(\nu)}$ is the character of the composite (product) representation.

This formula provides the Clebsch–Gordan coefficients purely at the level of characters.

13.3 Example: Decomposition of $E \otimes E$ in D_3

We illustrate the general procedure with a concrete example taken from the non-abelian group D_3.

From the character table we see that the character of the irrep $E = D^{(3)}$ of the group D_3 is

$$(2, -1, 0).$$

The character of the tensor product $E \otimes E$ is therefore

$$\chi^{(E \otimes E)} = (4,\ 1,\ 0),$$

since the character of a tensor product is the product of the characters,

$$\chi^{(E \otimes E)}(g) = \chi^{(E)}(g)\, \chi^{(E)}(g).$$

Hence, from the orthogonality relation we obtain

$$\text{For } D^{(1)}: \qquad a_1 = \frac{1}{6}\Big(1 \cdot 1 \cdot 4 + 2 \cdot 1 \cdot 1 + 3 \cdot 1 \cdot 0\Big) = 1,$$

$$\text{For } D^{(2)}: \qquad a_2 = \frac{1}{6}\Big(1 \cdot 1 \cdot 4 + 2 \cdot 1 \cdot 1 + 3 \cdot (-1) \cdot 0\Big) = 1,$$

$$\text{For } D^{(3)}: \qquad a_3 = \frac{1}{6}\Big(1 \cdot 2 \cdot 4 + 2 \cdot (-1) \cdot 1 + 3 \cdot 0 \cdot 0\Big) = 1.$$

where we used the character orthogonality relation

$$a_\nu = \frac{1}{|G|} \sum_i p_i \, \chi^{(E\otimes E)}(g_i) \, \chi^{(\nu)}(g_i)^*.$$

Therefore

$$\chi^{(E\otimes E)} = \chi_1 + \chi_2 + \chi_3,$$

and in crystallographic notation we obtain

$$E \otimes E = A_1 \oplus A_2 \oplus E.$$

This decomposition reflects the emergence of symmetric and antisymmetric combinations of the two-dimensional representation E.

13.4 Direct Product Representation: When It Is Irreducible

We now summarize the precise conditions under which tensor-product representations remain irreducible.

We mentioned above that two particles behave very differently with respect to spatial transformations and the symmetry thereof, depending on whether they are non-interacting (uncoupled) or interacting (coupled).

For two independent particles (non-interacting), the symmetry group of the whole systems comprising of the two particles, is the direct product $G \times G$. The key point is that irreducibility depends on the symmetry group, not on the representation alone.

Each particle transforms under its own copy of the group:

$$(g_1, g_2) \in G \times G.$$

Particle 1 transforms in the irreducible representation $D^{(\mu)}$, particle 2 in the irreducible representation $D^{(\nu)}$:

$$\Psi \longmapsto \left(D^{(\mu)}(g_1) \otimes D^{(\nu)}(g_2)\right) \Psi.$$

This is a representation of the direct-product group $G \times G$.

Every irreducible representation of $G \times G$ is of the form

$$D^{(\mu)} \otimes D^{(\nu)},$$

and such tensor products are irreducible.

Thus, $D^{(\mu)}(g_1) \otimes D^{(\nu)}(g_2)$ is an irreducible representation of $G \times G$.

If the two particles are bound by a strong interaction, instead, clearly they must undergo the same transformation $g \in G$, and the symmetry group is a single copy of G:

$$\Psi \longmapsto \left(D^{(\mu)}(g) \otimes D^{(\nu)}(g)\right) \Psi.$$

This defines a representation of G:

$$D(g) = D^{(\mu)}(g) \otimes D^{(\nu)}(g).$$

Such tensor-product representations are generally reducible and decompose as

$$D^{(\mu)} \otimes D^{(\nu)} = \bigoplus_{\sigma} a_\sigma \, D^{(\sigma)},$$

where a_σ are multiplicities and the $D^{(\sigma)}$ are irreducible representations of G. This reducibility encodes the internal coupling structure of the composite system.

Thus, in this case $D^{(\mu)}(g) \otimes D^{(\nu)}(g)$ is reducible as a representation of G, because its dimensionality vastly exceeds the dimension of any irrep of G.

The multiplicity a_σ of the irreducible representation $D^{(\sigma)}$ inside the tensor product $D^{(\mu)} \otimes D^{(\nu)}$ is given by

$$a_\sigma = \frac{1}{|G|} \sum_{g \in G} \chi^{(\mu)}(g) \, \chi^{(\nu)}(g) \, \chi^{(\sigma)}(g)^*.$$

Here,

$$\chi^{(\mu)}(g) \text{ is the character of irrep } \mu,$$

$$\chi^{(\nu)}(g) \text{ is the character of irrep } \nu,$$

$$\chi^{(\upsilon)}(g)^* \text{ is the complex conjugate of the character of irrep } \sigma,$$

and $|G|$ is the order of the group.

Hence, this formula computes how many times the irrep σ appears in the decomposition

$$D^{(\mu)} \otimes D^{(\nu)} = \bigoplus_{\sigma} a_\sigma \, D^{(\sigma)}.$$

Hence, the Clebsch–Gordan decomposition provides a complete and systematic method for reducing tensor product representations into irreducible components, thereby revealing the symmetry structure of composite systems.

13.5 Semidirect Product, Groups and Packings

We now move beyond direct products to a more subtle composition of symmetries, where one group acts non-trivially on another.

As a concrete illustration of how symmetry groups and their representations constrain geometric optimization problems, we consider the classical problem of packing congruent disks in the Euclidean plane. This example shows how group structure can dictate optimal geometric arrangements. This example also serves the purpose of introducing the concept of semidirect product, which is a generalization of the direct product or tensor product of groups discussed in the present chapter.

This example is discussed extensively in the monograph by Conway and Sloane, *Sphere Packings, Lattices and Groups* [3], and provides a natural bridge between abstract group theory and geometric applications.

13.5.1 Lattice Packings and Symmetry

A *lattice packing* in $\mathbb{R}^2$ is obtained by placing identical disks of radius r with centers at the points of a lattice $\Lambda \subset \mathbb{R}^2$. The lattice is generated by two linearly independent vectors $\mathbf{a}_1, \mathbf{a}_2$, and its fundamental cell has area

$$A = |\det(\mathbf{a}_1, \mathbf{a}_2)|.$$

The packing condition requires that the distance between any two distinct lattice points be at least $2r$. Denoting by $d_{\min}(\Lambda)$ the length of the shortest nonzero lattice vector, this constraint reads

$$d_{\min}(\Lambda) \geq 2r.$$

The packing density is defined as the fraction of area covered by disks:

$$\delta(\Lambda) = \frac{\pi r^2}{A}.$$

Maximizing the density is therefore equivalent to minimizing the fundamental area A under the constraint $d_{\min}(\Lambda) \geq 2r$. Symmetry considerations thus strongly restrict the possible minimizers.

13.5.2 Optimality Among Lattices

Proposition 13.1 *Among all lattice packings of congruent disks in* $\mathbb{R}^2$*, the maximal density is*

$$\delta_{\max} = \frac{\pi}{2\sqrt{3}},$$

and is achieved uniquely (up to rigid motions) by the hexagonal lattice.

Proof Let $\mathbf{a}_1$ and $\mathbf{a}_2$ be lattice basis vectors, and let θ be the angle between them. The area of the fundamental cell is

$$A = \|\mathbf{a}_1\| \, \|\mathbf{a}_2\| \sin\theta.$$

At the optimal point, both $\mathbf{a}_1$ and $\mathbf{a}_2$ may be chosen to be shortest lattice vectors, hence

$$\|\mathbf{a}_1\| = \|\mathbf{a}_2\| = 2r.$$

The vector $\mathbf{a}_1 - \mathbf{a}_2$ is also a nonzero lattice vector and must satisfy

$$\|\mathbf{a}_1 - \mathbf{a}_2\| \geq 2r.$$

Using the cosine law,

$$\|\mathbf{a}_1 - \mathbf{a}_2\|^2 = (2r)^2 + (2r)^2 - 2(2r)(2r)\cos\theta = 8r^2(1 - \cos\theta),$$

which implies

$$\cos\theta \leq \frac{1}{2}, \qquad \theta \geq \frac{\pi}{3}.$$

Therefore

$$A = 4r^2 \sin\theta \geq 4r^2 \sin\left(\frac{\pi}{3}\right) = 2\sqrt{3}\,r^2,$$

and the density satisfies

$$\delta(\Lambda) = \frac{\pi r^2}{A} \leq \frac{\pi}{2\sqrt{3}}.$$

Equality holds if and only if $\theta = \pi/3$, in which case the lattice vectors form equilateral triangles and the lattice is hexagonal. □

13.5.3 Symmetry Group of the Hexagonal Packing

The hexagonal lattice exhibits the largest possible point symmetry among planar lattices, which is a direct consequence of Theorem 7.1, that we proved in Sect. 7.3.1 of Chap. 7 (the only possible rotation angles for Bravais lattices are of order $n = 1, 2, 3, 4, 6$). Its full symmetry group is the wallpaper group

$$\Gamma = \Lambda \rtimes C_6,$$

where $\Lambda \cong \mathbb{Z}^2$ is the discrete translation group of the lattice and C_6 is the cyclic group generated by a rotation through 60° in the plane (and thus has the same properties of cyclic groups C_n introduced in Chap. 7, Sect. 7.4). Here, $\mathbb{Z}^2 = \{(m, n) \mid m, n \in \mathbb{Z}\}$ denotes the discrete translation group in 2D.

This structure generalizes the direct product by allowing one subgroup to act on the other.

The semidirect product $G = N \rtimes H$ describes a group in which a normal subgroup N (typically translations) is combined with a subgroup H (typically rotations or boosts) that acts nontrivially on N, so that the group law reflects the fact that transformations in H change the way elements of N are composed.

13.6 Semidirect Product

The semidirect product structure reflects the nontrivial action of rotations on translations:

$$R\, T_{\mathbf{a}}\, R^{-1} = T_{R\mathbf{a}}, \qquad R \in C_6,\ \mathbf{a} \in \Lambda.$$

This large symmetry group strongly restricts the geometry of the packing and explains why the hexagonal configuration naturally emerges as the extremal solution. The best way to understand the concept of semidirect group is via yet another example.

Example: The Two-Dimensional euclidean Group as a Semidirect Product
This example illustrates semidirect products in a familiar physical setting.

A simple and physically transparent example of a semidirect product is provided by the group of rigid motions of the Euclidean plane, the *two-dimensional Euclidean group* $E(2)$. This group consists of all distance-preserving transformations of $\mathbb{R}^2$, namely translations and rotations.

Let

$$N = \mathbb{R}^2 \equiv \{(x, y) \mid x, y \in \mathbb{R}\}$$

denote the abelian continuous group of translations in the plane, and

$$H = \mathrm{SO}(2)$$

the group of rotations about the origin. An element of $E(2)$ can be written as a pair $(\mathbf{a}, R)$, where $\mathbf{a} \in \mathbb{R}^2$ is a translation vector and $R \in \mathrm{SO}(2)$ is a rotation.

A translation by $\mathbf{a}$ acts on a point $\mathbf{x} \in \mathbb{R}^2$ as

$$T_{\mathbf{a}}: \quad \mathbf{x} \mapsto \mathbf{x} + \mathbf{a},$$

while a rotation acts as

$$R: \quad \mathbf{x} \mapsto R\,\mathbf{x}.$$

The crucial observation is that rotations act *nontrivially* on translations by conjugation. Indeed, for any $R \in \mathrm{SO}(2)$ and $\mathbf{a} \in \mathbb{R}^2$, one finds

$$R\,T_{\mathbf{a}}\,R^{-1} = T_{R\mathbf{a}}.$$

Thus, under a rotation, a translation vector $\mathbf{a}$ is mapped into the rotated vector $R\mathbf{a}$. In other words, rotations *relabel* translations by changing their direction in the plane.

This shows that the translation subgroup $N = \mathbb{R}^2$ is normal in $E(2)$, but that it is not central: translations commute among themselves, and rotations commute among themselves, but translations and rotations do not commute with each other. This failure of commutativity is precisely what prevents the group from being a direct product. Also, the failure of commutativity is precisely encoded by the action of $\mathrm{SO}(2)$ on $\mathbb{R}^2$.

As a result, the Euclidean group in two dimensions is not a direct product but a semidirect product,

$$E(2) = \mathbb{R}^2 \rtimes \mathrm{SO}(2).$$

The group law follows from the above action and is given by

$$(\mathbf{a}, R)\,(\mathbf{b}, S) = (\mathbf{a} + R\mathbf{b},\ RS).$$

This example illustrates concretely what it means for one group to act nontrivially on another: the structure of translations depends on the rotational frame in which they are described. Semidirect products therefore naturally arise whenever one set of symmetries transforms another, as is the case for spacetime symmetries, crystal symmetries, and the Poincaré group discussed later in Chap. 18.

13.6.1 Beyond Lattices

The above argument establishes optimality *among lattice packings*. A nontrivial geometric theorem, due originally to Thue, shows that no non-lattice packing in the plane can exceed this density. Consequently, the hexagonal packing is the densest possible packing of congruent disks in $\mathbb{R}^2$.

This result illustrates a recurring theme in mathematical physics and geometry: highly symmetric configurations are natural candidates for extremal solutions, and group-theoretic methods provide a powerful organizing principle for identifying and classifying them.

While here we presented a relatively simple example, the applications of group theory to determine the densest packings of spheres in a certain spatial dimension d is an important research direction in both pure mathematics and physics [4]. It is also the third part of Hilbert's 18th problem, which asks for the densest sphere packing or packing of other specified shapes. This problem, for 3D ordered packings, was fully solved only in recent years when Hales rigorously proved the Kepler conjecture (i.e. that the face centered cubic lattice, with about 0.74 volume fraction occupied by the spheres, is the densest possible arrangements of spheres in 3D) [5]. The corresponding problem for *random* sphere packings can be solved only within the framework of statistical physics, and a closed-form analytical solution was presented in [6] by using the theory of correlation functions in disordered media.

A spectacular modern example of the power of symmetry methods in packing problems is the exact solution of the sphere packing problem in eight dimensions, achieved by Viazovska using modular forms and harmonic analysis [7].

In general, tensor products and semidirect products together provide the mathematical framework needed to describe composite systems, coupled degrees of freedom, and spatial symmetries in both quantum mechanics and geometry.

References

1. H.F. Jones, *Groups, Representations and Physics*, 2nd edn. (Taylor & Francis Ltd, Bristol/New York, 1998)
2. J.F. Cornwell, *Group Theory in Physics*, vol. 1 (Academic Press, London, 1997)
3. J.H. Conway, N.J.A. Sloane, *Sphere packings, lattices and groups*. Grundlehren der mathematischen Wissenschaften, vol. 290, 3rd edn. (Springer, New York, 1999)
4. S. Torquato, F.H. Stillinger, Rev. Mod. Phys. **82**(3), 2633 (2010). https://doi.org/10.1103/RevModPhys.82.2633
5. T.C. Hales, Ann. Math. **162**(3), 1065 (2005)
6. A. Zaccone, Phys. Rev. Lett. **128**, 028002 (2022). https://doi.org/10.1103/PhysRevLett.128.028002. https://link.aps.org/doi/10.1103/PhysRevLett.128.028002
7. M. Viazovska, Ann. Math. **185**(3), 991 (2017)

Chapter 14
Physical Properties of Solids and Energy-Level Splitting in Quantum Mechanics

Abstract In this chapter we apply the principles derived in the previous chapters to obtain qualitative information about physical systems. We will start with condensed matter physics, and in particular with ferroelectricity, ferromagnetism and the electric conductivity of solids. We then move to nonrelativistic quantum mechanics and atomic physics where historically the first application of group theory has been to rationalize the energy level splitting of atoms under external electrical or magnetic fields.

14.1 Physical Properties of Crystalline Solids

14.1.1 Permanent Electric Dipole and Magnetic Dipole Moments

Given a crystal with a certain symmetry, can it have a permanent magnetic dipole moment **M** or a permanent electric dipole moment **P**? The answer to these questions can be obtained from group theory by knowing nothing else than the point group symmetry of the crystal. The fundamental condition for a crystal to possess these properties is that such quantities (**M**, **P**) be invariant with respect to the transformations of the crystal point group G, which consists of rotations and reflections (the translations of the lattice, which transform the crystal into itself, are not relevant here). This is because the physical property of interest here, is a pseudo-vector with cartesian components:

$$\mathbf{M} = (M_x, M_y, M_z).$$

for the case of the magnetic dipole moment, and a standard vector with cartesian components:

$$\mathbf{P} = (P_x, P_y, P_z).$$

for the case of the electric dipole moment of the crystal.

A. Zaccone, *Differential Geometry and Group Theory*,
https://doi.org/10.1007/978-3-032-27518-9_14

The fundamental requirement is invariance under the crystal point group P. According to Neumann's principle, a physical property may be nonzero only if it is invariant under all symmetry operations of the crystal.

In representation-theoretic language, this means that the vector representation D^V must contain the trivial representation A_1 with nonzero multiplicity. Only then can a nonzero invariant vector exist. Importantly, **M** is a vector which transforms under rotations according to the three-dimensional vector representation:

$$M_i' = D_{ij}^V(g)\, M_j .$$

for proper rotations, but it transforms, instead, as a pseudo-vector under improper rotations, i.e. picking up an extra -1 factor under reflection.

Let us first consider the case in which P does not contain reflections, but is a subgroup of the full rotation group.

In order for a dipole moment to exist that is invariant under the action of P, it is necessary that

$$\mathbf{M}' = \mathbf{M} \qquad \forall g \in P.$$

The fundamental principle here is that physical properties cannot change when one applies a symmetry operation that leaves the crystal invariant. Therefore, the representation D^V, when restricted to P, must contain the identity representation.

In order for this to be possible, the crystal must possess an invariant axis. For example, in a crystal with trigonal symmetry there exists a well-defined direction along which the moments **M** or **P** can point.

Let us now see how all this is translated into the language of characters. For C_3 we have calculated the decomposition of D^V:

$$D^V = D^{(1)} \oplus D^{(2)} \oplus D^{(3)}.$$

In particular, the identity representation $D^{(1)}$ occurs only once, which means that there exists a unique direction along which **M** can point.

If the symmetry is higher, in the sense that there is a larger number of symmetry transformations, this may no longer be the case. For example, in D_3 there is no longer an invariant spatial direction. From a physical point of view, C_3 is a cyclic group describing the symmetry of a triangle with oriented sides, and therefore there is a distinction between the $+z$ direction (counterclockwise sense of rotation about the z axis) and the $-z$ direction. In other words, each side of the triangle also carries a label related to the sense of rotation. This distinction is lost in D_3 because of the presence of dihedral rotations, which exchange the directions (clockwise and counterclockwise) of the sides. The existence of dihedral rotations is in fact responsible for the decomposition of the vector representation D^V for the transformations that are applied to a three-dimensional vector, which in the case of D_3 is:

$$D^V = A_2 \oplus E,$$

and therefore it does not contain the trivial identity representation $A_1 \equiv D^{(1)}$.

It is important to bear in mind the difference between **P** and **M**: indeed, they have a different behavior under the parity operation (or spatial inversion), and hence under improper rotations. In particular, **P** follows the behavior of the electric field **E**:

$$\mathbf{P} \to -\mathbf{P} \qquad \text{when} \qquad \mathbf{x} \to -\mathbf{x}.$$

On the other hand, **M**, like **B**, is an axial pseudo-vector, that is, it is associated with an axis rather than a direction, and when passing from **x** to −**x** it remains invariant. This is different from the behavior of **P**, which is a polar vector, and it changes sign depending on the orientation (up or down) along the axis. If the spatial transformations of a certain symmetry group include reflections or inversions or improper rotations, then this already restricts its ability to support a permanent electric moment.

A crystal supports a permanent dipole moment if and only if the corresponding vector quantity is not forced to vanish by the symmetry operations of its point group.

An electric dipole moment **P** transforms as a polar vector, whereas a magnetic dipole moment **M** transforms as an axial (pseudovector). The distinction between polar and axial vectors becomes relevant in the presence of improper symmetry operations.

Under inversion I, a polar vector changes sign,

$$\mathbf{P} \xrightarrow{I} -\mathbf{P},$$

while an axial vector remains invariant,

$$\mathbf{M} \xrightarrow{I} \mathbf{M}.$$

As a consequence, any point group containing inversion symmetry forbids a permanent electric dipole moment, since invariance would require $\mathbf{P} = -\mathbf{P}$, implying $\mathbf{P} = 0$. In contrast, inversion does not forbid a permanent magnetic dipole moment.

Improper rotations S_n combine a proper rotation with a reflection and therefore reverse the handedness of space. Because of this, polar and axial vectors transform differently under such operations, and strong constraints are imposed on the allowed components of physical vectors.

In particular, inversion symmetry forbids a permanent electric dipole moment, while mirror planes generally restrict the allowed directions of both polar and axial vectors. The precise constraints depend on how the vector representation decomposes into irreducible representations of the point group.

If a point group contains only proper rotations and no inversion, mirror planes, or improper rotations, then both polar and axial vectors may remain invariant along certain directions. Such groups can therefore support both permanent electric and permanent magnetic dipole moments.

These results are summarized in the following table.

Symmetry present	Electric dipole **P**	Magnetic dipole **M**
Inversion (I)	Forbidden	Allowed
Mirror plane (σ)	Generally forbidden	Generally forbidden
Improper rotation (S_n)	Forbidden	Forbidden
Only proper rotations	Allowed	Allowed

Consequently, the symmetry elements of a crystal place strict constraints on which components of physical vectors and tensors (such as polarization, magnetization, or elasticity tensors) can be nonzero. In practice, this means that a physical property can exist only if it transforms according to the totally symmetric representation of the crystal's point group; any component that is not invariant under the symmetry operations must vanish.

We thus see that symmetry alone—without microscopic modeling—determines whether a crystal can exhibit ferroelectricity or ferromagnetism. This is one of the most powerful practical applications of group theory in condensed matter physics.

Exercise

The octahedral group O_h is the full symmetry group of a regular octahedron (or equivalently of a cube). It contains all symmetry operations that leave the octahedron invariant, including both proper and improper operations, and has a total order of 48.

The subgroup of proper rotations is denoted by O and contains 24 elements. These include the identity operation, three fourfold rotation axes C_4 passing through the centers of opposite faces of the cube, four threefold rotation axes C_3 passing through opposite vertices, and six twofold rotation axes C_2 passing through the midpoints of opposite edges. Together, these rotations generate all orientation-preserving symmetries of the octahedron.

The full group O_h, also denoted as T, is obtained by adding inversion symmetry to the rotation group O. The inversion operation i maps each point (x, y, z) to $(-x, -y, -z)$ and makes the group centrosymmetric. For every proper rotation $R \in O$, there exists a corresponding improper operation iR, so that the set of improper operations has the same cardinality as the set of proper ones.

As a consequence, O_h contains mirror planes and improper rotations such as S_4 and S_6, in addition to inversion. The presence of these improper symmetry elements implies that the group is nonpolar and achiral. Using a similar reasoning, determine

whether the octahedral group O_h can support a permanent electric dipole moment or a permanent magnetic dipole moment or both.
Answer: Because O_h contains inversion, mirror planes, and improper rotations, it forbids both a permanent electric dipole moment and a permanent magnetic dipole moment. Only physical properties that are invariant under all 48 symmetry operations of the group can be nonzero, in accordance with Neumann's principle.

14.2 Conductivity Tensor of Crystals

Vector quantities are the simplest example of symmetry-constrained physical observables. More generally, many measurable properties of solids are described by tensors.

The electrical conductivity provides a fundamental example of a rank-2 tensor whose allowed form is determined entirely by symmetry. In an anisotropic solid such as a crystal, the electric field **E** and the current density **j** are not simply proportional, but can be directed along different orientations. In this case, the general relation between the two vectors is given by a tensor σ_{ij} with two indices (rank 2):

$$j_i = \sigma_{ij} E_j.$$

The tensor σ transforms like the (direct) product of the components of two vectors.

In rotated coordinates:

$$j_i' = \sigma_{ij}' E_j'.$$

Using vector-matrix notation,

$$\mathbf{j}' = D^V \mathbf{j}, \qquad \mathbf{E}' = D^V \mathbf{E}.$$

Substituting $\mathbf{j} = \sigma \mathbf{E}$ and multiplying on the left by D^V,

$$\mathbf{j}' = D^V \sigma (D^V)^{-1} \mathbf{E}'.$$

Comparing with $\mathbf{j}' = \sigma' \mathbf{E}'$ gives

$$\sigma' = D^V \sigma (D^V)^{-1}.$$

Since D^V is orthogonal, $(D^V)^{-1} = (D^V)^T$, and therefore

$$\sigma' = D^V \sigma (D^V)^T.$$

In components this reads

$$\sigma'_{ij} = D^V_{ik}\,\sigma_{kl}\,D^V_{jl}.$$

Thus the conductivity tensor transforms according to the tensor product representation

$$V \otimes V.$$

We are therefore facing the same problem encountered earlier: which components of this reducible representation transform as the totally symmetric representation? Only those components can remain invariant under the crystal symmetry. We therefore ask: how many times does the trivial representation A_1 appear in $(V \otimes V)_+$? This multiplicity equals the number of independent invariant tensor components.

Every rank-2 tensor can be decomposed into a symmetric and an antisymmetric part:

$$T_{ij} = T^{(s)}_{ij} + T^{(a)}_{ij},$$

with

$$T^{(s)}_{ij} = \frac{1}{2}(T_{ij} + T_{ji}), \qquad T^{(a)}_{ij} = \frac{1}{2}(T_{ij} - T_{ji}),$$

which belong to two distinct invariant subspaces. Under rotations, they transform independently of each other:

$$T'_{ij} = D^V_{ik}\,D^V_{jl}\,T_{kl}$$

$$T^{(s)\prime}_{ij} = \frac{1}{2}\left[D^V_{ik}\,D^V_{jl} + D^V_{jk}\,D^V_{il}\right]T_{kl}$$

$$= \frac{1}{2}\left(\underbrace{D^V_{ik}\,D^V_{jl}\,T_{kl}}_{T'_{ij}} + \underbrace{D^V_{jk}\,D^V_{il}\,T_{kl}}_{T'_{ji}}\right)$$

and analogously for the antisymmetric part. Thus, upon collecting, we get:

$$\Rightarrow \quad D^{(V\otimes V)}_{ij,\,kl} = \frac{1}{2}\left[D^V_{ik}\,D^V_{jl} \pm D^V_{jk}\,D^V_{il}\right].$$

The characters are obtained by contraction with $\delta_{ik}\delta_{jl}$:

$$\chi^{(V\otimes V)_\pm}(g) = \frac{1}{2}\left[\left(\chi^V(g)\right)^2 \pm \chi^V(g^2)\right].$$

$$= \frac{1}{2}\left[\left(\chi^V(g)\right)^2 \pm \chi^V(g^2)\right].$$

In the last step we have used:

$$D^V_{ij}(g)\, D^V_{ji}(g) = \chi^V(g^2).$$

In plain words, this is the matrix product of the same representation (same matrix representation of the same group element), and therefore gives the trace of the matrix of the representation of the squared element, $\chi^V(g^2)$. Clearly, this is tantamount to using the homomorphism, i.e. the group composition law.

Let us illustrate this principle with the examples studied in the previous chapter. We already computed the decomposition of the vector representation for the groups C_3 and D_3. We can now reinterpret those results physically.

Example: Direct Product Characters of D_3

Using the results that we deduced in the previous section, in the case of group D_3 we obtain:

	e	(c, c^2)	(b, bc, bc^2)
V	3	0	-1
$(V\otimes V)_+$	6	0	2
$(V\otimes V)_-$	3	0	-1

For the identity, e, we have:

$$(V\otimes V)_+ = \frac{1}{2}\left[\left(\chi^V(e)\right)^2 + \chi^V(e^2)\right] = \frac{1}{2}[9+3] = 6 \qquad \text{(symmetric)}$$

$$(V\otimes V)_- = \frac{1}{2}[9-3] = 3 \qquad \text{(antisymmetric)}$$

In the case of b we have:

$$(V\otimes V)_+ = \frac{1}{2}\left[\left(\chi^V(b)\right)^2 + \chi^V(b^2)\right] = \frac{1}{2}[(-1)(-1)+3] = 2$$

remembering that

$$b^2 = e \quad \Rightarrow \quad \chi^V(b^2) = \chi^V(e).$$

From an intuitive point of view: in the tensor

$$T_{ij}^{(s)} = \frac{1}{2}(T_{ij} + T_{ji})$$

there are 6 independent elements, whereas in the tensor

$$T_{ij}^{(a)} = \frac{1}{2}(T_{ij} - T_{ji})$$

there are only 3, since the diagonal elements are all zero.

Therefore, the restrictive condition on the possible forms of the conductivity tensor σ_{ij} is that these must be invariant under the symmetry transformations of the symmetric direct vector product representation:

$$(V \otimes V)_+ \qquad (+ \text{ since } \sigma_{ij} \text{ is symmetric})$$

for all elements of the crystallographic point group P, $\forall\, g \in P$.

Let us consider in detail the case where the crystallographic point group of the solid is D_3.

The simplest form that we always have for the conductivity tensor is

$$\sigma_{ij} = \delta_{ij},$$

which is trivial, since it is invariant under all rotations (the invariance of δ_{ij} is a manifestation of the fact that the matrices D^V are orthogonal).

The problem now is whether there exist other forms that are invariant with respect to the rotation subgroup of D_3.

To solve the problem we must use the character table and ask: how many times does the representation $(V \otimes V)_+$ contain the identity representation A_1? This multiplicity equals the number of independent invariant tensor components. In practice, as usual, this implies computing the coefficient a_1 of the Clebsch-Gordan decomposition of the reducible representation $(V \otimes V)_+$ into the trivial identity irrep A_1,

$$a_1 = \langle \chi^{(1)}, \chi^{(V \otimes V)_+} \rangle = \frac{1}{6}(1 \cdot 1 \cdot 6 + 2 \cdot 1 \cdot 0 + 3 \cdot 1 \cdot 2) = 2$$

using, as always,

$$a_\nu = \frac{1}{|G|} \sum_i p_i \left[\chi^{(\nu)}(g_i)\right]^* \chi(g_i).$$

Hence, because $a_1 = 2$, from our calculation above, this means that there is another basic form of the 3 x 3 matrix σ_{ij}, besides the Kronecker delta δ_{ij}, that is invariant under the symmetry operations for the direct vector product representation of the group (D_3, in our example).

This other form of the matrix is the diagonal matrix

$$\mathrm{diag}(0, 0, 1),$$

or $\delta_{i3}\delta_{j3}$, since it is immediately seen that by applying a rotation to this tensor it remains invariant under all transformations of D_3.

Therefore, the most general form of the matrix σ_{ij} is a linear combination of the two forms that we have found, and the fact that there are no other forms other than these two is guaranteed by group theory through the study of the irreps of $(V \otimes V)_+$ carried out above.

Hence, any linear combination of the two basic invariant forms

$$\sigma = \begin{pmatrix} a & 0 & 0 \\ 0 & a & 0 \\ 0 & 0 & b \end{pmatrix},$$

that is, a linear combination of

$$\begin{pmatrix} 1 & 0 & 0 \\ 0 & 1 & 0 \\ 0 & 0 & 1 \end{pmatrix} \quad \text{and} \quad \begin{pmatrix} 0 & 0 & 0 \\ 0 & 0 & 0 \\ 0 & 0 & 1 \end{pmatrix}.$$

will provide a suitable form of the electrical conductivity tensor σ_{ij} for crystals of the group D_3.

Example: Cubic Crystals

Let us now consider a crystal with cubic symmetry (simple cubic lattice, described by the octahedral group O_h). The character table is as follows:

	E	$3C_2$	$4C_3$	$4C_3^2$
V	3	-1	0	0
$(V \otimes V)_+$	6	2	0	0

From this we obtain:

$$a_1 = \frac{1}{12}(1 \cdot 6 + 3 \cdot 2 + 4 \cdot 0 + 4 \cdot 0) = 1,$$

which means that $(V \otimes V)_+$ contains the identity representation only once, and only one invariant form of σ_{ij} exists. Again, this form must coincide with the Kronecker delta δ_{ij} because this is the only form invariant under all spatial rotations.

This implies that the only possible form of the conductivity tensor is:

$$\sigma_{ij} = \sigma\,\delta_{ij}.$$

This is a manifestation of the fact that the simple cubic lattice is macroscopically isotropic (it exhibits the same properties in all cartesian directions). Strictly speaking, the simple cubic lattice is not isotropic microscopically, because it is invariant only under the finite cubic point group (O_h or T) and not under the full rotation group SO(3). However, the cubic symmetry is high enough that any second-rank tensor invariant under the crystal symmetries, such as the conductivity or dielectric tensor, is forced to be proportional to δ_{ij}. As a result, these physical properties appear the same in all directions and are therefore macroscopically isotropic. This emergent isotropy is not fundamental but symmetry-enforced, and it does not generally extend to higher-rank tensors, where anisotropy can still occur.

14.2.1 Higher-Rank Tensors of Physical Properties: The Elastic Tensor

The same reasoning extends to higher-rank tensors. In elasticity theory, stress and strain are rank-2 tensors, and the elastic constants form a rank-4 tensor. Symmetry severely restricts the number of independent elastic constants in a crystal. Rank-n tensors transform as

$$V \otimes V \otimes \cdots \otimes V \qquad (n \text{ times}),$$

and, for example, in the theory of elasticity of crystals [1]:

$$\sigma_{kl} = \lambda_{klmn}\,\varepsilon_{mn},$$

where σ_{kl} is, now, the stress tensor, while ε_{mn} is the strain tensor, both of which are rank-2 tensors. The elastic constants tensor λ_{klmn} is a 4th-rank tensor and therefore it transforms according to the following triple direct product of the vector representation:

$$\lambda_{klmn} \;\Rightarrow\; V \otimes V \otimes V \otimes V.$$

14.3 Level Splitting and Symmetry-Breaking in Quantum Mechanics

So far we have applied group theory to constrain static material properties. We now turn to a conceptually different but mathematically analogous application: the role of symmetry in quantum mechanics, and in particular in the degeneracy and splitting of energy levels.

In the remaining of this chapter we shall study in detail what consequences the Clebsch-Gordan decomposition of a reducible representation into irreps bears for quantum mechanics, and in particular for the energy levels of atoms.

14.3.1 Degeneracy of Energy Levels in Quantum Mechanics

In general, the stationary Schrödinger equation can be written as

$$H\psi^{\alpha} = E^{\alpha}\psi^{\alpha} \qquad \text{(no summation convention on } \alpha\text{)},$$

where α labels the eigenvectors ψ^{α} of the quantum Hamiltonian H, corresponding to its eigenvalues E^{α}, which represent the energy levels of the system. It may happen that there exists a subset of eigenfunctions

$$\psi^{\alpha}, \qquad \alpha = 1, \ldots, r,$$

such that

$$H\psi^{\alpha} = E\psi^{\alpha} \qquad \forall\alpha = 1, \ldots, r.$$

In practice, all the eigenfunctions ψ^{α} share the same eigenvalue E, which implies a degeneracy of the energy levels.

In general we have an *unperturbed* quantum Hamiltonian, which in bra-ket notation reads as:

$$H_0\,|E^{(0)}\rangle = E^{(0)}\,|E^{(0)}\rangle.$$

with $|E^{(0)}\rangle$ the eigenstates of H_0 and $E^{(0)}$ the corresponding eigenvalues. Suppose that a set of transformations $U(g)$ (realized by unitary operators) leaves the Hamiltonian invariant:

$$U^{\dagger}(g)\,H_0\,U(g) = H_0.$$

Since U is unitary,

$$U^{\dagger} = U^{-1},$$

and therefore the invariance condition can be expressed by the vanishing of a commutation bracket as

$$[\,H_0,\;U(g)\,] = 0.$$

Here $U(g)$ is the unitary operator that acts in Hilbert space and represents the symmetry transformation applied to the system. Thanks to the invariance of H_0 we obtain:

$$(H_0 U)\,|E^{(0)}\rangle = U H_0\,|E^{(0)}\rangle$$

where the commutation relation has been used, leading to:

$$H_0\big(U|E^{(0)}\rangle\big) = U E^{(0)}|E^{(0)}\rangle = E^{(0)}\big(U|E^{(0)}\rangle\big).$$

Therefore, the group transformation $U(g)$, when applied to an eigenstate $|E^{(0)}\rangle$ of the Hamiltonian H_0, produces another eigenstate of the same Hamiltonian, with the same eigenvalue $E^{(0)}$. In other words, $U(g)$ transforms eigenstates into one another. These eigenstates therefore form a subspace within the full vector space of eigenstates of H_0 and provide a basis in Hilbert space.

If there is only one eigenstate $E^{(0)}$ in the subspace, the representation reduces to the trivial one (of dimension one) and the energy level is non-degenerate because there is only one eigenvector $|E^{(0)}\rangle$ associated with the eigenvalue $E^{(0)}$. If there is more than one eigenstate, however, we have degenerate energy levels, because the eigenstates share the same eigenvalue $E^{(0)}$. In this case, the action of the group induces a r-dimensional representation, where r is the number of degenerate eigenstates.

In general, this representation will be irreducible, because there is no reason to expect the existence of smaller invariant subspaces.

Therefore, an energy level $E^{(0)}$ that has a degeneracy equal to 3, for example, corresponds to an irreducible representation $D^{(\nu)}$ of G with

$$r = \dim D^{(\nu)} = 3.$$

If there are other degeneracies (besides the one just described), one speaks of *accidental degeneracy*, which may be due to an additional, "hidden" symmetry.

Therefore, a given energy level $E^{(0)}$ corresponds to a given irreducible representation $D^{(\nu)}$ of the group G, and the degeneracy r (that is, the number of eigenfunctions that share the same eigenvalue $E^{(0)}$) is equal to the dimensionality n_μ of the irrep.

All this can also be viewed from the point of view of the first Schur lemma, as follows. Consider an energy eigenspace $\mathcal{H}_E$ of a Hamiltonian H_0 with eigenvalue $E^{(0)}$. Since

$$[H_0, U(g)] = 0 \qquad \forall g \in G,$$

the space $\mathcal{H}_E$ is invariant under the action of the symmetry group G. Therefore, $\mathcal{H}_E$ carries a representation D of G. Now, assume that this representation is irreducible.

Then, by Schur's first lemma (cf. Chap. 11), any operator A acting on $\mathcal{H}_E$ that commutes with all representation matrices,

$$[A, D(g)] = 0 \qquad \forall g \in G,$$

must be proportional to the identity:

$$A = \lambda\,\mathbf{1}.$$

Applying this result to the Hamiltonian itself, since H_0 commutes with all $U(g)$ and the representation of G on $\mathcal{H}_E$ is irreducible, Schur's lemma implies

$$H_0\big|_{\mathcal{H}_E} = E^{(0)}\,\mathbf{1}.$$

Thus degeneracy is not an accident: it is enforced by symmetry whenever the eigenstates form an irreducible representation of dimension greater than one. This means that all states belonging to the same irreducible representation have the same energy, and therefore the degeneracy of the level is exactly equal to the dimension of the irrep.

Conversely, if an energy level is degenerate but the degeneracy cannot be accounted for by the dimensionality of an irreducible representation of the symmetry group, then the representation is reducible or incomplete. In this case one speaks of *accidental degeneracy*, which must be due to an additional (hidden) symmetry.

In conclusion, symmetry-enforced degeneracy arises because the Hamiltonian acts as a scalar operator on irreducible representation spaces, and this result follows directly from Schur's lemma.

Example: Hidden Degeneracy in the Hydrogen Atom

Let us consider the energy spectrum of a particle moving in a central potential. The symmetry group of the system is SO(3). Its irreducible representations have dimension $2\ell + 1$ and are labeled by the angular momentum quantum number ℓ, which is associated with the angular part of the wavefunction. The magnetic quantum number

$$m = -\ell, \ldots, +\ell$$

labels the states within a given irrep and accounts for the $2\ell + 1$ degeneracy, which is directly enforced by rotational symmetry.

The principal quantum number n is instead associated with the radial part of the wavefunction.

For a generic central potential $V(r)$, the energy depends on both n and ℓ:

$$E = E_{n\ell},$$

and the only degeneracy is the $(2\ell+1)$-fold degeneracy in m, which is explained by the SO(3) symmetry. This is a standard degeneracy, and is *not* a hidden degeneracy.

For the Coulomb potential

$$V(r) = -\frac{k}{r},$$

the situation is different: the energy depends only on the principal quantum number n,

$$E = E_n, \qquad 0 \le \ell \le n-1.$$

As a consequence, states with different values of ℓ have the same energy.

This additional degeneracy in ℓ is *not* explained by the SO(3) symmetry and is therefore called *accidental (or hidden) degeneracy* from the SO(3) point of view.

For a fixed n, the total degeneracy is

$$\sum_{\ell=0}^{n-1}(2\ell+1) = n^2.$$

This means that an energy level with fixed n corresponds to a reducible representation of SO(3), since it contains several irreducible representations with different values of ℓ.

The origin of this degeneracy lies in the fact that the Coulomb Hamiltonian possesses a larger hidden symmetry group, SO(4), generated by the angular momentum and the Runge–Lenz vector. Under SO(4), all states with the same principal quantum number n belong to a single irreducible representation of dimension n^2 [2].

This example illustrates that degeneracy patterns reveal hidden symmetry groups. Conversely, observing degeneracies experimentally can guide the identification of the underlying symmetry.

14.3.2 Lifting of Degeneracy (Level-Splitting)

Let us now assume that we introduce an external perturbation V that does not respect the symmetry of the Hamiltonian H_0:

$$H_1 = H_0 + V,$$

but is invariant under a subgroup

$$H \subset G,$$

that is, the system now has a lower (or reduced) symmetry, compared to the unperturbed system described by H_0.

When a perturbation reduces the symmetry, the relevant mathematical operation is restriction of representations from G to a subgroup $H \subset G$.

The energy levels of the perturbed Hamiltonian can be classified by the irreducible representations of H instead of those of G. Therefore, when we consider a given energy level, what was previously an irrep of G must now be decomposed into irreps of H. One should note that an irreducible representation of G, in general, does *not* remain irreducible when it is restricted to a subgroup $H \subset G$, because irreducibility means that there are no invariant subspaces under the action of *all* elements $g \in G$. When restricting to a subgroup, one considers only the elements

$$g \in H,$$

and therefore invariant subspaces may appear that were not invariant under the full group. Indeed, an irreducible representation of a group G is defined by the absence of non-trivial invariant subspaces under the action of *all* elements of G. When the representation is restricted to a subgroup $H \subset G$, invariance is required only with respect to a smaller set of elements. Since fewer constraints are imposed, subspaces that were not invariant under the full group G may become invariant under H. As a consequence, a representation that is irreducible for G can become reducible when restricted to a subgroup.

In particular, an irrep μ of G may decompose into a direct sum of irreducible representations of H labelled by κ:

$$D^{(\mu)} \longrightarrow \bigoplus_{\kappa} D^{(\kappa)}.$$

Physically, this means that the degeneracy associated with the original symmetry G is partially or completely lifted when the symmetry is reduced to the subgroup H.

Let us consider the case of the energy levels of an atom, which is first considered in its unperturbed state and then is placed in a perturbing external field. For example, this could be a magnetic field (Zeeman effect) or an electric field (Stark effect), a situation schematically depicted in Fig. 14.1.

The quantities $E_{\alpha}^{(0)}$ and $E_{\beta}^{(0)}$ are eigenvalues (energy levels) of the unperturbed Hamiltonian H_0, corresponding, respectively, to irreducible representations labeled by α and β (indices of the irreps). On the other hand, $E_{\alpha\rho}^{(1)}$ and $E_{\alpha\sigma}^{(1)}$ are eigenvalues of H_1, whose irreducible representations are labeled by ρ and σ, respectively.

In practice, by using the character tables, we can determine the irreducible representations of H into which the irreducible representations of G decompose upon introducing the perturbation, and therefore reconstruct the pattern of the level splitting. This process can be best understood with a concrete calculation example, which is provided in the next subsection.

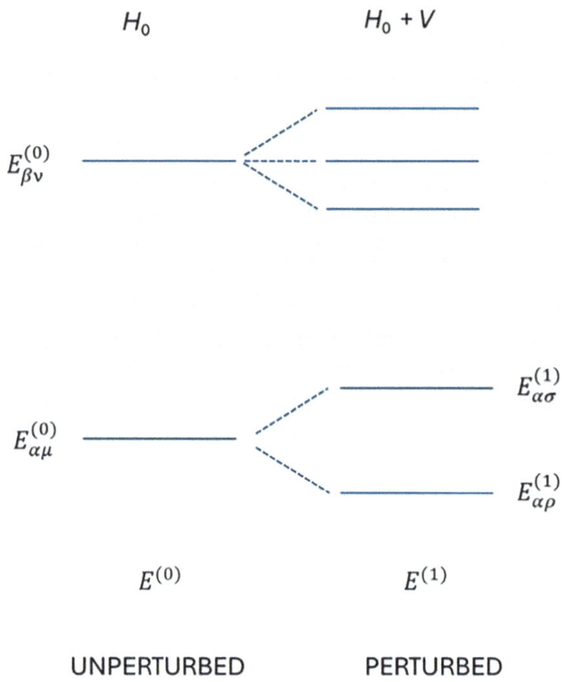

Fig. 14.1 Schematic splitting of degenerate energy levels under symmetry reduction. On the left, the unperturbed Hamiltonian H_0 has degenerate energy levels $E^{(0)}_{\beta\nu}$ and $E^{(0)}_{\alpha\mu}$, associated with irreducible representations of the full symmetry group G, ν and μ, respectively. After the introduction of an external perturbation V, the perturbed Hamiltonian $H_1 = H_0 + V$ is invariant only under a subgroup $H \subset G$, and the degeneracies are partially lifted. Each irreducible representation of G decomposes into irreducible representations of H, producing the split energy levels $E^{(1)}_{\alpha\rho}$ and $E^{(1)}_{\alpha\sigma}$. The number and pattern of split levels are determined by the breaking of symmetry from G to H

Example: Crystal Field Splitting (Hans Bethe, 1929)
If we insert an atom (for example Na, Cu or Al) into a crystalline solid, the electrons of the atom will feel the Coulomb potential generated by the ions of the crystal lattice. In practice, the original, unperturbed Hamiltonian of the atom is what we called H_0, while the perturbed Hamiltonian of the atom placed inside the crystal lattice is $H_1 = H_0 + V$, where here V is a very complicated field which receives contributions from the other atoms that form the crystal lattice. In most solids, such as in metals, the atoms of the lattice are ionized or they bear a net electric charge, thus they are ions. The sum of all the electrostatic interactions with the periodically arranged ions of the lattice leads to the perturbation potential V. Since the ions are arranged in the lattice according to the point symmetry group P of the crystal, the symmetry that was the SO(3) symmetry of the free atom in empty space, is now broken or reduced to the point-group symmetry P of the crystal.

As a consequence, energy levels that were degenerate in the free atom split when the atom is inserted into the crystal, with a splitting pattern (like those shown in Fig. 14.1) determined by the decomposition of the reducible SO(3) representation within the environment of the discrete, and lower-symmetry, point group P.

Let us study the example of a Cu atom inserted into a cubic lattice with a crystallographic structure given by the octahedral group.

The character table for the group O_h, also denoted as T, is given by:

T	E	$3C_2$	$4C_3$	$4C_3^2$
A	1	1	1	1
E_1	1	1	ω	ω^2
E_2	1	1	ω^2	ω
T	3	-1	0	0

where

$$\omega = e^{2\pi i/3}.$$

Here:

- $3C_2$ are rotations by π about the axes passing through the midpoints of opposite edges of the cube;
- $4C_3$ are rotations by $2\pi/3$ about the four body diagonals of the cube;
- $4C_3^2$ are rotations by $4\pi/3$ about the same diagonals.

We anticipate an important result concerning the characters of SO(3) for orbital angular momentum ℓ:

$$\chi^{(\ell)}(g) = \frac{\sin\left[(\ell + \frac{1}{2})\theta\right]}{\sin(\frac{1}{2}\theta)} = 1 + 2\big(\cos\theta + \cos 2\theta + \cdots + \cos \ell\theta\big).$$

Let us consider, for example, a d orbital, corresponding to $\ell = 2$. This is meaningful because the external electronic structure of copper (Cu) is indeed of d-type. We thus compute the character of the SO(3) representation for the rotations of the group T, namely for

$$\theta = 0,\ \pi,\ \frac{2\pi}{3},\ \frac{4\pi}{3}.$$

Thus,

$$\chi^{(2)}(\theta) = (5,\ -1,\ -1,\ -1),$$

where the value 5 corresponds to the identity.

Therefore, using the orthogonality relations, we have

$$a_\nu = \frac{1}{|G|} \sum_i p_i \, \chi^{i^{(\ell)}} \left[\chi^{i^{(\nu)}}\right]^* = \langle \chi^{(\ell)}, \chi^{(\nu)} \rangle.$$

$$a_A = \frac{1}{12} \left(1 \cdot 5 \cdot 1 + 3 \cdot 1 \cdot 1 + 4 \cdot (-1) \cdot 1 + 4 \cdot (-1) \cdot 1\right) = 0$$

$$\begin{aligned} a_{E_1} &= \frac{1}{12}\Big(1 \cdot 5 \cdot 1 + 3 \cdot 1 \cdot 1 + 4 \cdot \mathrm{Re}\Big(e^{-i2\pi/3}\Big)(-1) \\ &\quad + 4 \cdot \mathrm{Re}\Big(e^{-i4\pi/3}\Big)(-1)\Big) = 1 \end{aligned}$$

$$\begin{aligned} a_{E_2} &= \frac{1}{12}\Big(1 \cdot 5 \cdot 1 + 3 \cdot 1 \cdot 1 + 4 \cdot \mathrm{Re}\Big(e^{-i4\pi/3}\Big)(-1) \\ &\quad + 4 \cdot \mathrm{Re}\Big(e^{-i2\pi/3}\Big)(-1)\Big) = 1 \\ a_T &= \frac{1}{12}\left(1 \cdot 5 \cdot 3 + 3 \cdot (-1) \cdot (-1) + 0 + 0\right) = 1 \\ &\Rightarrow \qquad D^{(\ell=2)} = E_1 \oplus E_2 \oplus T. \end{aligned}$$

Initially, we had a single d-wave energy state, which is five-fold degenerate

$$m = -2, -1, 0, +1, +2,$$

but the anisotropic field of the O_h crystal lattice, in which the atom is inserted, breaks the SO(3) symmetry of empty space. In particular, the representation of dimension 5, becomes now reducible and decomposes into irreducible representations of the tetrahedral O_h group:

$$5 \longrightarrow 1 + 1 + 3.$$

which we can write as

$$D^{(2)}_{SO(3)}\Big|_T = D^{(E_1)} \oplus D^{(E_2)} \oplus D^{(T)}.$$

or in full crystallographic notation:

$$D^{(\ell=2)}\big|_T = E_1 \oplus E_2 \oplus T.$$

Thus the fivefold degeneracy of the d orbitals splits into two singlets and one triplet under cubic symmetry. The precise energy ordering depends on the details of the crystal field, but the splitting pattern itself is dictated purely by symmetry.

Exercise

We have seen that, in order for a crystal to possess a permanent electric dipole moment **P** or a permanent magnetic moment **M**, it is necessary that the vector representation (acting on a vector (x, y, z)) of the crystal point group G, when decomposed into irreducible representations, contains the identity representation.

If the decomposition contains the identity representation A_1 once, for example, then there exists a single invariant axis, and therefore a unique direction along which **P** or **M** can point.

When the point group G contains only *proper rotations*, the vector representation is the same for polar and axial vectors, and the above criterion applies equally to **P** and **M**.

If the point group contains *improper operations* (reflections or inversion), polar and axial vectors transform differently, and the decomposition must be analyzed separately for **P** and **M**.

We have seen that the group C_3, which contains only proper rotations, has the identity representation appearing once in the vector representation, so that $a_{A_1} = 1$ and the crystal can support a permanent electric or magnetic dipole moment.

On the other hand, the group D_3, which also contains only proper rotations, has $a_{A_1} = 0$ in the vector representation and therefore cannot support a permanent dipole moment.

Let us now consider the group C_{3v} of the ammonia molecule. This group contains improper operations (reflections), and is isomorphic as an abstract group to D_3,

$$C_{3v} \simeq D_3,$$

but it is *not* equivalent as a point group because the presence of reflections changes the transformation properties of polar and axial vectors.

The symmetry operations of C_{3v} correspond to those of a right pyramid with an equilateral triangular base, as depicted in Fig. 14.2. The group is generated by 3-fold rotations about the axis OD and by reflections σ_v through the plane OAD.

(i) Show that

$$C_{3v} \simeq D_3.$$

(ii) Construct the dim $= 3$ vector representation acting on a polar vector

$$\mathbf{v} = (x, y, z).$$

NB: the matrix for b is *not* the same as the one we saw for D_3.

(iii) Show that a crystal with this symmetry can possess a permanent electric dipole moment **P**.

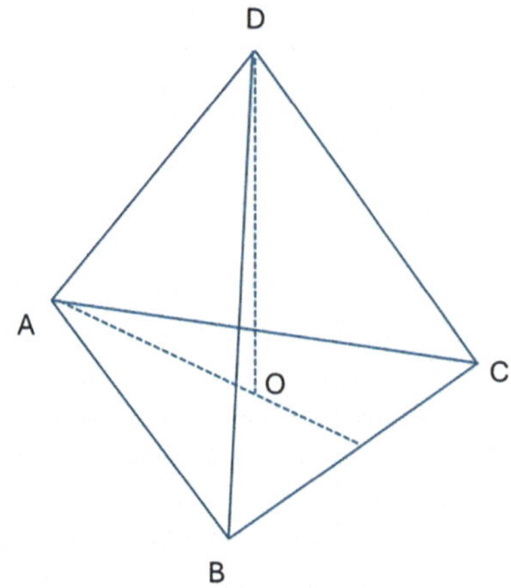

Fig. 14.2 Right pyramid on an equilateral triangle basis

(iv) **M**, like the magnetic field **B**, is an axial vector; therefore, under a reflection, it transforms with the opposite sign compared to a polar vector. Show that a crystal with C_{3v} symmetry cannot support a permanent magnetic dipole moment.

Solution
(i) Show that $C_{3v} \simeq D_3$.
Let c be the 3-fold rotation about the principal axis and let $b = \sigma_v$ be a vertical reflection. Then

$$c^3 = e, \qquad b^2 = e.$$

Moreover, the product bc is again a reflection (about a different vertical plane), hence it has order 2:

$$(bc)^2 = e.$$

Therefore C_{3v} has the elements

$$\{c, b \mid c^3 = e,\ b^2 = e,\ (bc)^2 = e\},$$

which is the standard presentation of the dihedral group of order 6, i.e. D_3. Hence

$$C_{3v} \simeq D_3.$$

Intuition: both groups describe "symmetries of an equilateral triangle": the rotations $\{e, c, c^2\}$ and three order-2 elements ("flips").
(ii) Construct the dim $= 3$ *(vector) representation on a polar vector* $\mathbf{v} = (x, y, z)$.

Choose coordinates so that the 3-fold axis is the z-axis, and choose the reflection plane σ_v to be the xz-plane (so $y \mapsto -y$). Then

$$D(c) = \begin{pmatrix} \cos\frac{2\pi}{3} & -\sin\frac{2\pi}{3} & 0 \\ \sin\frac{2\pi}{3} & \cos\frac{2\pi}{3} & 0 \\ 0 & 0 & 1 \end{pmatrix}, \qquad D(b) = \begin{pmatrix} 1 & 0 & 0 \\ 0 & -1 & 0 \\ 0 & 0 & 1 \end{pmatrix}.$$

Intuition: a polar vector transforms exactly like the coordinates under rotations and reflections.

(iii) Show that a crystal with C_{3v} symmetry can have a permanent electric dipole moment **P**.

A permanent electric dipole **P** is a *polar* vector. It can exist only if there is a nonzero vector invariant under all symmetry operations:

$$D(g)\mathbf{P} = \mathbf{P} \qquad \forall g \in C_{3v}.$$

Equivalently, the polar-vector representation V must contain the trivial irrep A_1. Using characters on the conjugacy classes $(E,\ 2C_3,\ 3\sigma_v)$,

$$\chi^V(E) = 3, \qquad \chi^V(C_3) = 2\cos\frac{2\pi}{3} + 1 = 0, \qquad \chi^V(\sigma_v) = \mathrm{tr}\, D(b) = 1,$$

hence

$$\chi^V = (3,\ 0,\ 1), \qquad \chi^{A_1} = (1,\ 1,\ 1), \qquad |C_{3v}| = 6.$$

The multiplicity of A_1 in V is

$$a_{A_1} = \frac{1}{6}\Big(1 \cdot 3 \cdot 1 + 2 \cdot 0 \cdot 1 + 3 \cdot 1 \cdot 1\Big) = 1.$$

Therefore V contains A_1 once, so there is a one-dimensional invariant direction.

Intuition: the only direction left unchanged by both the C_3 rotation and all σ_v mirrors is the principal axis; thus **P** is allowed only along that axis.

(iv) Show that a crystal with C_{3v} symmetry cannot have a permanent magnetic dipole moment **M**.

A magnetic dipole **M** is an *axial* vector (pseudovector), like **B**. Under proper rotations axial and polar vectors transform the same way, but under reflections they differ by an extra sign (because an axial vector behaves like a cross product of two polar vectors). Thus the axial-vector characters on $(E,\ 2C_3,\ 3\sigma_v)$ are

$$\chi^{\mathrm{ax}}(E) = 3, \qquad \chi^{\mathrm{ax}}(C_3) = 0, \qquad \chi^{\mathrm{ax}}(\sigma_v) = -1,$$

i.e.

$$\chi^{\text{ax}} = (3,\ 0,\ -1).$$

The multiplicity of A_1 is then

$$a_{A_1} = \frac{1}{6}\Big(1 \cdot 3 \cdot 1 + 2 \cdot 0 \cdot 1 + 3 \cdot (-1) \cdot 1\Big) = \frac{1}{6}(3-3) = 0.$$

Hence the axial-vector representation contains no trivial component, so there is no nonzero invariant $\mathbf{M}$: a C_{3v} crystal cannot support a permanent magnetic dipole moment. *Intuition:* the mirror planes force any would-be axial vector to flip sign under reflection, so symmetry requires $\mathbf{M} = 0$.

References

1. L.D. Landau, E.M. Lifshitz, *Theory of Elasticity*. Course of Theoretical Physics, vol. 7, 3rd edn. (Butterworth-Heinemann, Oxford, 1986)
2. L.D. Landau, E.M. Lifshitz, *Quantum Mechanics: Non-relativistic Theory*. Course of Theoretical Physics, vol. 3, 3rd edn. (Pergamon Press, Oxford, 1977)

Chapter 15
Representation Theory of SO(2) and SO(3)

Abstract In this chapter we delve into the full representation theory of the most important Lie rotation groups in physics, SO(2) and SO(3). In the previous chapters we developed the representation theory of finite groups. We now extend these ideas to continuous Lie groups, where the role of discrete sums over group elements is replaced by integrals with respect to an invariant (Haar) measure, and where the structure of the group is encoded in its Lie algebra of infinitesimal generators.

15.1 Continuous Rotation Groups

Continuous rotations by arbitrary angles in $2, 3, \ldots, N$ dimensions form compact groups, in which the angular parameters vary over finite intervals, such as $[0, \pi]$ or $[0, 2\pi]$.

We can adapt the character method that we developed for discrete groups, with the caveat of replacing the sum over group elements

$$\frac{1}{|G|}\sum_{g}\cdots$$

by an appropriate integral equipped with a suitable integration measure.

It is also useful to recall the Lie algebra of the infinitesimal generators of the group (presented in Chap. 9). The Lie algebra $\mathfrak{so}(3)$ is generated by three operators X_i satisfying the commutation relations

$$[X_i, X_j] = i\varepsilon_{ijk}X_k.$$

In quantum mechanics these generators are realized by the angular momentum operators $J_i = \hbar X_i$, which therefore satisfy

$$[J_i, J_j] = i\hbar\,\varepsilon_{ijk}J_k.$$

A. Zaccone, *Differential Geometry and Group Theory*,
https://doi.org/10.1007/978-3-032-27518-9_15

It is precisely by exploiting the formal equivalence between the Lie algebra of SO(3) and the algebra of the angular momentum in quantum mechanics that it will be possible to derive the irreducible representations of SO(3), later in this chapter.

15.2 The Group SO(2)

The group SO(2) is the simplest nontrivial example of a compact Lie group. Being abelian, its representation theory is particularly simple and provides a prototype for Fourier analysis.

15.2.1 Irreps and Characters

Rotations about the z-axis in the xy-plane.

A rotation by an angle φ is represented by

$$R(\varphi) = \begin{pmatrix} \cos\varphi & -\sin\varphi \\ \sin\varphi & \cos\varphi \end{pmatrix}, \tag{15.1}$$

which gives the defining (vector) representation of the group.

The kernel contains only the identity element ($\varphi = 0$). The vector representation [2] (dim $= 2$) is irreducible over $\mathbb{R}$, but it becomes reducible over $\mathbb{C}$.

Indeed, introducing the complex combinations

$$\begin{pmatrix} x + iy \\ x - iy \end{pmatrix},$$

the rotation acts diagonally:

$$\begin{pmatrix} x + iy \\ x - iy \end{pmatrix} \longrightarrow \begin{pmatrix} e^{i\varphi} & 0 \\ 0 & e^{-i\varphi} \end{pmatrix} \begin{pmatrix} x + iy \\ x - iy \end{pmatrix},$$

which is tantamount to diagonalizing the rotation matrix in (15.1).

Equivalently, if we write

$$z = x + iy = re^{i\theta},$$

then under a rotation by φ we have

$$z \longrightarrow z' = re^{i(\theta+\varphi)},$$

with

$$r' = r, \qquad \theta' = \theta + \varphi.$$

We therefore obtain faithful one-dimensional representations, given by 1×1 matrices (i.e. just numbers in the complex field $\mathbb{C}$), which define the one-dimensional irreducible representations of SO(2) (equivalently of the isomorphic group $U(1)$).

Finally, the abelian group law is reflected in the representation through

$$R(\varphi)\, R(\varphi') = R(\varphi + \varphi').$$

Hence, the irreps of SO(2) in $\mathbb{C}$ are all one-dimensional.

We also have

$$D^{(m)}(\varphi) = e^{-im\varphi}, \qquad m \in \mathbb{Z}.$$

The sign convention in the exponential is a matter of definition and depends on whether one writes the representation as acting on functions or on basis states; both conventions are equivalent.

The group SO(2) is compact and is parametrized by a continuous angle φ with $0 \leq \varphi < 2\pi$. Therefore, the discrete average over the group elements,

$$\frac{1}{|G|} \sum_g \ldots,$$

can be replaced by a continuous integral

$$\frac{1}{2\pi} \int_0^{2\pi} d\varphi \ldots,$$

which yields the correct orthonormality relations for the characters. The latter statement can be easily verified.

Indeed,

$$\langle \chi^{(m)}, \chi^{(m')} \rangle = \int_0^{2\pi} \frac{d\varphi}{2\pi}\, e^{im\varphi}\, e^{-im'\varphi} = \delta_{mm'}.$$

The Clebsch–Gordan series for the decomposition of the product of two irreducible representations is given by

$$D^{(m)} \otimes D^{(m')} = D^{(m+m')}.$$

15.2.2 Decomposition of the Representation

Let us consider again the rotation matrix in Eq. (15.1) and let us decompose it into irreducible representations in the m-basis.

The multiplicity a_m of the irreducible representation $D^{(m)}$ is given by the scalar product of characters:

$$a_m = \langle \chi^{(m)}, \chi^{(R)} \rangle = \frac{1}{2\pi} \int_0^{2\pi} d\varphi \, e^{-im\varphi} \, \chi^{(R)}(\varphi).$$

For the vector representation R, the character is

$$\chi^{(R)}(\varphi) = \mathrm{Tr}\, R(\varphi) = 2\cos\varphi.$$

Thus

$$a_m = \frac{1}{2\pi} \int_0^{2\pi} d\varphi \, e^{-im\varphi} \, 2\cos\varphi = \frac{1}{2\pi} \int_0^{2\pi} d\varphi \, e^{-im\varphi} \left(e^{i\varphi} + e^{-i\varphi} \right).$$

Evaluating the integral, one finds

$$a_m = \delta_{m,1} + \delta_{m,-1}.$$

Therefore the decomposition of the representation is

$$R = D^{(1)} \oplus D^{(-1)}.$$

The basis of the irreducible representations is given by

$$x \pm iy.$$

15.2.3 Infinitesimal Generators

To study the Lie algebra of the generators of SO(2), one can determine them by expanding a generic rotation about the origin:

$$\mathsf{R}(\varphi) = \mathbf{1} - i\varphi \mathsf{X} + o(\varphi^2) \quad \Rightarrow \quad \mathsf{X} = i \left. \frac{d\mathsf{R}(\varphi)}{d\varphi} \right|_{\varphi=0} = \begin{bmatrix} 0 & -i \\ i & 0 \end{bmatrix} \tag{15.2}$$

This matrix X happens to coincide with the Pauli matrix σ_2, although here it arises as the generator of planar rotations rather than as part of the SU(2) algebra.

The fact that the infinitesimal generator of SO(2) is Hermitian (in particular, a Pauli matrix) follows from the orthogonality condition:

$$\mathbf{1} = \mathsf{R}(\varphi)^\dagger \mathsf{R}(\varphi) = \left(\mathbf{1} + i\varphi \mathsf{X}^\dagger\right)(\mathbf{1} - i\varphi \mathsf{X}) = \mathbf{1} - i\varphi\left(\mathsf{X} - \mathsf{X}^\dagger\right) + o(\varphi^2)$$

$$\Rightarrow \mathsf{X}^\dagger = \mathsf{X}.$$

Moreover, since rotation matrices have unit determinant, it follows that the infinitesimal generator must be traceless.

For proper rotations (continuously connected to the identity, and therefore excluding reflections), one has:

$$\mathsf{R}(\varphi) = \exp(-i\varphi \mathsf{X}) . \tag{15.3}$$

This proposition can be easily verified by recalling that $\sigma_2^2 = \mathbf{1}$, we have:

$$\exp(-i\varphi \mathsf{X}) = \sum_{k=0}^{\infty} \frac{(-i\varphi)^k}{k!} \mathsf{X}^k = \mathbf{1}\cos\varphi - i\mathsf{X}\sin\varphi = \mathsf{R}(\varphi).$$

For a complex irreducible representation ρ_m, instead, the generator is trivial:

$$\mathsf{X}_m = -m.$$

So far we have discussed rotations as abstract matrices acting on vectors. We now reinterpret these rotations as unitary operators acting on quantum states.

15.3 Connection with Quantum Mechanics

In the context of quantum mechanics, the infinitesimal generator is an operator $\hat{X}$ acting on the wave function $\psi(\mathbf{x})$. Note that, after a rotation in physical space, the rotated wave function must retain the same value at the rotated point:

$$\psi'(\mathbf{x}') = \psi(\mathbf{x}),$$

as illustrated in Fig. 15.1 for the rotation of a p-orbital wavefunction.

As a consequence, a rotation R defines the unitary operator that transforms the wave function, $\hat{U}_\mathsf{R}$, as

$$\hat{U}_\mathsf{R}\psi(\mathbf{x}) = \psi(\mathsf{R}^{-1}\mathbf{x}). \tag{15.4}$$

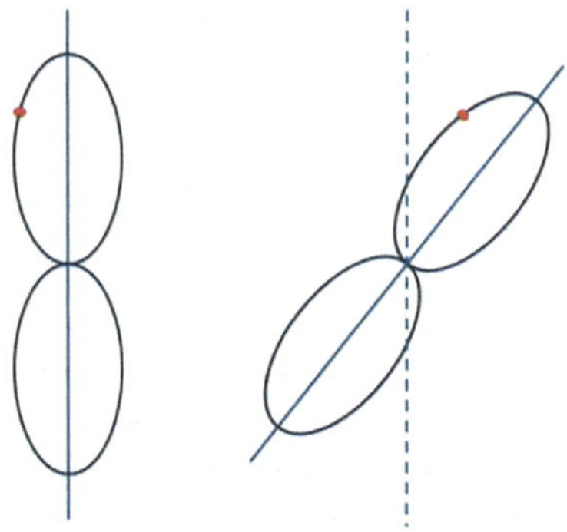

Fig. 15.1 Geometric illustration of a rotation acting on a quantum state. The wave function is represented schematically by a shape whose value at a given point (marked by the red dot) is preserved under rotation: after the rotation, the value of the wave function at the rotated point equals the original value at the original point, $\psi'(\mathbf{x}') = \psi(\mathbf{x})$. This illustrates why the rotation operator acts as $\hat{U}_{\mathsf{R}}\psi(\mathbf{x}) = \psi(\mathsf{R}^{-1}\mathbf{x})$ and motivates the appearance of the angular derivative as the infinitesimal generator

Assuming for simplicity azimuthal symmetry and expanding about the origin, one finds

$$(1 - i\varphi\hat{X} + o(\varphi^2))\,\psi(r,\theta) = \psi(r,\theta-\varphi) \quad \Rightarrow \quad i\hat{X} = \frac{\partial}{\partial\theta}.$$

By rescaling $\hat{X}$ by a factor $\hbar$, one recovers the usual generator of the angular momentum along the z-axis:

$$\hat{J}_z = -i\hbar\frac{\partial}{\partial\theta}. \tag{15.5}$$

The irreducible representations ρ_m of SO(2) admit basis eigenfunctions $u_m(\theta) = e^{im\theta}$, which satisfy

$$\hat{X}u_m = m\,u_m.$$

Indeed,

$$\hat{X}\,e^{im\varphi} = -i\frac{d}{d\varphi}e^{im\varphi} = m\,e^{im\varphi}.$$

From this one obtains the eigenvalue equation

$$\hat{J}_z u_m = \hbar m\,u_m. \tag{15.6}$$

which defines the angular momentum operator for a 2D, in-plane, rotation about the cartesian axis z.

15.3.1 What Are the Eigenfunctions of $\hat{J}_z$?

Let us take a closer look at the eigenfunctions of the rotation operator in 2D for an in-plane rotation by an angle φ about the z-axis which is orthogonal to the 2D plane:

$$\hat{X}\, u_m = m\, u_m$$

with eigenfunctions:

$$u_m(\varphi) = e^{im\varphi}$$

which coincide with the azimuthal-angle dependent part of the spherical harmonics $Y(\varphi, \theta)$:

$$Y_\ell^m(\theta, \varphi) = (-1)^m \sqrt{\frac{2\ell+1}{4\pi}\frac{(\ell-m)!}{(\ell+m)!}}\, P_\ell^m(\cos\theta)\, e^{im\varphi},$$

where $\ell = 0, 1, 2, \ldots$ and $m = -\ell, -\ell+1, \ldots, \ell$ while $P_\ell^m(x)$ are the associated Legendre functions [1].

The generator can be written as

$$\hat{X} = -\, i\, \frac{d}{d\varphi} = \frac{\hat{J}_z}{\hbar}.$$

Finite rotations are obtained by exponentiation:

$$R(\varphi) = e^{-\, i\varphi \hat{X}} = e^{-\, i\varphi m},$$

which represents finite rotations in two dimensions.

The generator $\hat{X}^{(m)}$ acts diagonally:

$$\hat{X}\, u_m = \hat{X}^{(m)} u_m, \qquad \hat{J}_z\, u_m = \hbar m\, u_m.$$

In three dimensions, the generators must satisfy commutation relations:

$$R_{\mathbf{n}}(\varphi) = e^{-\, i\, \mathbf{n}\cdot\mathbf{X}\, \varphi},$$

with

$$[X_1, X_2] = iX_3,$$

and cyclic permutations.

Finally, the action of a rotation on a function is given by

$$U(R)\,\psi(\mathbf{x}) = \sum_{mm'} D^{\ell}_{mm'}(R)\,\psi_{m'}(\mathbf{x}),$$

which represents the rotation in function space.

15.4 Mechanics of Rotations About an Arbitrary Axis

The group of all proper rotations in three dimensions is implemented by 3×3 matrices $R_{\hat{\mathbf{n}}}(\varphi)$, which describe rotations by an angle φ in the plane orthogonal to the vector (axis) $\hat{\mathbf{n}}$.

We have already studied the particular case

$$\hat{\mathbf{n}} = (0, 0, 1), \tag{15.7}$$

in which the matrices form a faithful representation of the abstract group. The matrices R are orthogonal, so as to preserve the scalar product, and they have determinant $+1$, like the identity matrix, to which they are continuously connected.

Let us again consider rotations about the z-axis:

$$R_z(\varphi) = \begin{pmatrix} \cos\varphi & -\sin\varphi & 0 \\ \sin\varphi & \cos\varphi & 0 \\ 0 & 0 & 1 \end{pmatrix}. \tag{15.8}$$

We compute the corresponding generator:

$$-iX_3 = \left.\frac{dR_z(\varphi)}{d\varphi}\right|_{\varphi=0} = \begin{pmatrix} 0 & -1 & 0 \\ 1 & 0 & 0 \\ 0 & 0 & 0 \end{pmatrix}. \tag{15.9}$$

Equivalently, in index notation,

$$(X_3)_{ij} = -i\varepsilon_{ij3}. \tag{15.10}$$

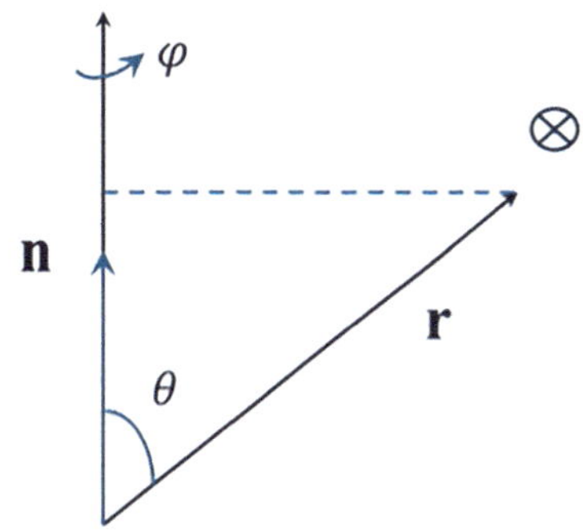

Fig. 15.2 Infinitesimal rotation of the vector **r** by an angle φ about the axis **n**. The change $\delta\mathbf{r}$ lies in the plane orthogonal to **n** and is given by $\delta\mathbf{r} = \varphi\,(\mathbf{n} \times \mathbf{r})$. The angle θ denotes the inclination of **r** with respect to the rotation axis

We have therefore

$$(X_1)_{ij} = -i\varepsilon_{ij1} \qquad \Rightarrow \qquad (X_k)_{ij} = -i\varepsilon_{ijk},$$

which represents a rotation about the k-th axis.

We can also visualize what happens in the following way, with reference to Fig. 15.2. With reference to the figure, consider a vector **n** and a rotation by an infinitesimal angle φ about an axis $\hat{\mathbf{n}}$. The displacement due to a small angle φ is given by the plane orthogonal to **n** and **r** and is given by $(R\sin\theta)\,\varphi$.

$$\delta\mathbf{r} = (\mathbf{n} \times \mathbf{r})\,\varphi \tag{15.11}$$

or equivalently

$$\mathbf{r}' = \mathbf{r} + \varphi\,(\mathbf{n} \times \mathbf{r}).$$

In components:

$$r_i' = r_i + \varphi\,\varepsilon_{ijk}\,n_j r_k = r_i - \varphi\,\varepsilon_{ijk}\,n_k r_j$$

(using the properties of the Levi–Civita symbol).

This can be rewritten as

$$r_i' = \big(\delta_{ij} - \varphi\,\varepsilon_{ijk} n_k\big) r_j.$$

In components:

$$r_i' = \big[R_{\mathbf{n}}(\varphi)\big]_{ij}\, r_j.$$

Therefore

$$\big[R_{\mathbf{n}}(\varphi)\big]_{ij} = \delta_{ij} - \varphi\,\varepsilon_{ijk} n_k = \delta_{ij} - i\varphi (X_{\mathbf{n}})_{ij},$$

where

$$(X_{\mathbf{n}})_{ij} = -i\varepsilon_{ijk}n_k = n_k(X_k)_{ij}.$$

Hence, the generator depends on the choice of the rotation axis **n**.

$$(X_k)_{ij} = -i\,\varepsilon_{ijk},$$

hence, given a direction **n** in space and the set of basis generators

$$X_1,\ X_2,\ X_3,$$

we can construct the generator of rotations about the axis **n**.

Define

$$\mathbf{X} = (X_1, X_2, X_3),$$

which can be regarded as a three-component vector in the space of generators. The generator associated with the direction **n** is then

$$X_{\mathbf{n}} = \mathbf{n} \cdot \mathbf{X}.$$

A generic rotation by an angle φ about the axis **n** is represented by

$$R_{\mathbf{n}}(\varphi) = e^{-i\,\varphi\,\mathbf{n}\cdot\mathbf{X}}.$$

Note The generator $X_{\mathbf{n}} = \mathbf{n} \cdot \mathbf{X}$ is purely imaginary and antisymmetric. As a consequence, the exponential

$$R_{\mathbf{n}}(\varphi)$$

is an orthogonal matrix with determinant equal to 1, i.e.

$$R_{\mathbf{n}}(\varphi) \in \mathrm{SO}(3).$$

15.4.1 Commutation Relations Revisited

A group is specified by its law of composition, that is, by how two rotations combine to form a single rotation. The structure of the group is encoded in the infinitesimal generators and in an associated algebra, namely a vector space equipped with a basis given by the generators X_i.

The algebra is specified by the commutation relations. In the algebraic approach, we have already found that

$$[X_i, X_j] = i\,\varepsilon_{ijk}\,X_k.$$

These relations can also be derived by considering how infinitesimal rotations combine, i.e. through the analysis of conjugated rotations.

Consider, for example, the transformation

$$R_{\mathbf{n}}(\varphi) \longrightarrow S(\theta)\,R_{\mathbf{n}}(\varphi)\,S^{-1}(\theta).$$

The group structure identifies the conjugated rotation as a rotation by the same angle φ, but about a rotated axis:

$$\mathbf{n}' = S\,\mathbf{n}.$$

Thus, conjugation corresponds to rotating the axis of the rotation while leaving the rotation angle unchanged.

Example: Rotation of the Axis

Let us choose the axis $\mathbf{n}$ along the x-direction, and let $S(\theta)$ be a rotation about the z-axis. The situation is depicted in Fig. 15.3.

Its representation is

$$S(\theta) = \exp(-iX_3\theta).$$

Under this rotation, the axis $\mathbf{n}$ is mapped to

$$\mathbf{n}' = S\,\mathbf{n},$$

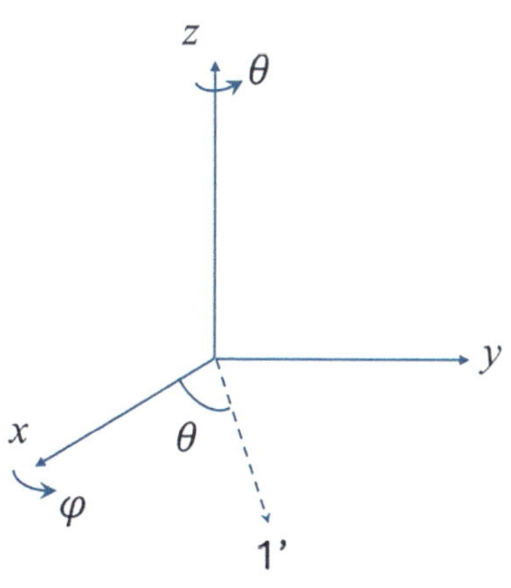

Fig. 15.3 Physical meaning of conjugation in SO(3): the transformation $S\,R_{\mathbf{n}}(\varphi)\,S^{-1}$ represents a rotation by the same angle φ, but about an axis $\mathbf{n}'$ which is given by $\mathbf{n}$ (originally aligned with the x-axis) that has been rotated by S

whose Cartesian components are

$$\mathbf{n}' = (\cos\theta,\ \sin\theta,\ 0).$$

Consider the conjugation

$$S(\theta)\, R_{\mathbf{n}}(\varphi)\, S^{-1}(\theta).$$

Using the exponential form of a finite rotation by φ around $\mathbf{n}$,

$$R_{\mathbf{n}}(\varphi) = e^{-i\,\mathbf{n}\cdot\mathbf{X}\,\varphi},$$

this becomes

$$S(\theta)\, R_{\mathbf{n}}(\varphi)\, S^{-1}(\theta) = e^{-iX_3\theta}\, e^{-iX_1\varphi}\, e^{iX_3\theta}.$$

Since the rotated axis is

$$\mathbf{n}' = (\cos\theta,\ \sin\theta,\ 0),$$

we expect

$$e^{-iX_3\theta}\, e^{-iX_1\varphi}\, e^{iX_3\theta} = e^{-i(X_1\cos\theta + X_2\sin\theta)\,\varphi}.$$

where we used

$$e^{-\,i\,\mathbf{n}\cdot\mathbf{X}\varphi} \;=\; e^{-\,i\,\left(X_1\cos\theta + X_2\sin\theta\right)\,\varphi}$$

and the fact that the z-component of $\mathbf{n}$ is zero.

Expanding to first order in φ, we find

$$e^{-iX_3\theta}\,(1 - iX_1\varphi)\, e^{iX_3\theta} = 1 - i\,\varphi\,\left(e^{-iX_3\theta}\, X_1 e^{iX_3\theta}\right).$$

We notice that the terms 1, as well as the factors $-i$ and φ, drop out on both sides, thus leaving:

$$e^{-iX_3\theta}\, X_1 e^{iX_3\theta} = X_1\cos\theta + X_2\sin\theta.$$

Differentiating the last identity with respect to θ and then taking the limit $\theta \to 0$, we finally obtain

$$-\,i[X_3, X_1] = X_2,$$

which is a defining commutation relation of SO(3). This derivation of the Lie commutation relation of SO(3) demonstrates an important principle: The commutation relations of the Lie algebra encode how infinitesimal rotations combine with one another by means of conjugation.

As we already point out several times, conjugacy classes correspond to all rotations with the same rotation angle, but with different axes that can be made to coincide with a rotation. The conjugated rotations are physically equivalent and can therefore be labeled solely by the angle φ.

The commutation relations thus have a clear geometrical meaning: they encode how infinitesimal rotations combine with one another.

We now turn to the central result of this chapter: the classification of the irreducible representations of SO(3). This can be achieved by exploiting the isomorphism between its Lie algebra and the algebra of angular momentum operators in quantum mechanics.

15.5 Irreps and Characters of SO(3)

We shall now embark on the derivation of the irreducible representations and characters of SO(3). We shall start by finding the irreps making use of the formal equivalence between the elements of SO(3) and the angular momentum operators of quantum mechanics.

15.5.1 Finding the Irreps of SO(3)

Conjugacy classes contain all rotations with the same rotation angle but about different axes. Therefore, they can be labeled solely by the rotation angle φ. In the same way, characters depend only on φ.

In the three-dimensional representation of $SO(3)$, one finds

$$\chi^{(3)}(\varphi) = 2\cos\varphi + 1,$$

as can be verified explicitly for rotations about the x-, y-, or z-axis.

The problem of finding the irreducible representations of the group is equivalent to finding irreducible matrices that satisfy the commutation relations

$$[X_1, X_2] = iX_3, \qquad \text{etc.}$$

What does it mean to represent the algebra? It means finding three matrices such that the commutation relations of the corresponding Lie algebra are satisfied. Since finite rotations are obtained by exponentiating the generators of the Lie algebra, this leads directly to matrix representations of the rotation group. Given Eq. (9.15), for connected Lie groups such as SO(3), the problem of finding irreducible representations of the group reduces to finding irreducible representations of its Lie algebra.

As already noted, the generators of SO(3) can be identified with the angular momentum operators in quantum mechanics (up to a factor $\hbar$), which satisfy the same commutation relations; indeed, from Eq. (15.11) in Sect. 15.4:

$$\psi'(\mathbf{x}) = \psi(\mathsf{R}_{\mathbf{n}}^{-1}\mathbf{x}) = \psi(\mathbf{x} - \delta\mathbf{x}) = \psi(\mathbf{x} - \phi\mathbf{n} \times \mathbf{x}) \tag{15.12}$$

$$= (1 - \phi(\mathbf{n} \times \mathbf{x}) \cdot \nabla)\, \psi(\mathbf{x}) = \left(1 - \frac{i}{\hbar}\phi\mathbf{n} \cdot \hat{\mathbf{J}}\right) \psi(\mathbf{x}). \tag{15.13}$$

In order to find the irreducible representations of SO(3) it is therefore possible to diagonalize these operators: the degree of the representation will be given by the number of eigenvalues thus obtained.

One notices that the operators $\hat{J}_k$ are not compatible (they do not commute); therefore, the following operators are defined:

$$\hat{J}_\pm \equiv \hat{J}_x \pm i\hat{J}_y \tag{15.14}$$

which satisfy the commutation relations:

$$[\hat{J}_\pm, \hat{J}^2] = 0$$

$$[\hat{J}_z, \hat{J}_\pm] = \pm\hbar\hat{J}_\pm$$

It is therefore possible to define a basis $|\beta, m\rangle$ such that:

$$\hat{J}^2|\beta, m\rangle = \beta^2\hbar^2|\beta, m\rangle$$

$$\hat{J}_z|\beta, m\rangle = m\hbar|\beta, m\rangle$$

The operators $\hat{J}_\pm$ can be interpreted as ladder operators:

$$\hat{J}_z\hat{J}_\pm|\beta, m\rangle = \left(\hat{J}_\pm\hat{J}_z + [\hat{J}_z, \hat{J}_\pm]\right)|\beta, m\rangle = \left(\hat{J}_\pm m\hbar \pm \hbar\hat{J}_\pm\right)|\beta, m\rangle$$

$$= (m \pm 1)\hbar\hat{J}_\pm|\beta, m\rangle$$

and their action is schematically depicted in Fig. 15.4.

Thus one must have $\hat{J}_\pm|\beta, m\rangle \propto |\beta, m \pm 1\rangle$. It can also be shown that the ladder thus constructed is bounded:

$$\beta^2\hbar^2 = \langle\beta, m|\hat{J}^2|\beta, m\rangle = \langle\beta, m|(\hat{J}_x^2 + \hat{J}_y^2 + m^2\hbar^2)|\beta, m\rangle \geq m^2\hbar^2$$

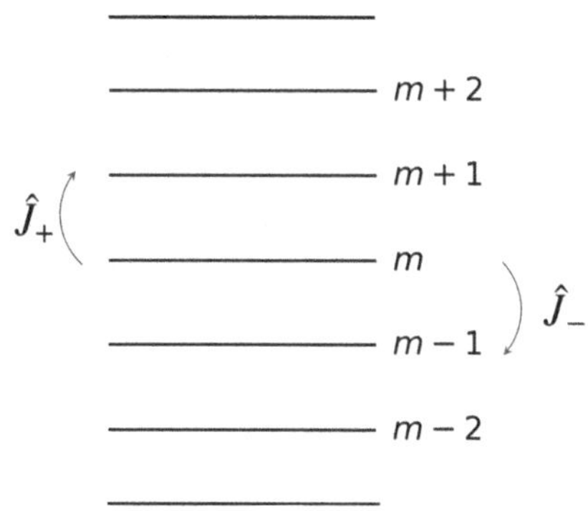

Fig. 15.4 Action of the ladder operators on angular momentum eigenstates. Starting from a state with magnetic quantum number m, repeated application of $\hat{J}_+$ or $\hat{J}_-$ generates a finite ladder of equally spaced states with m differing by integer steps

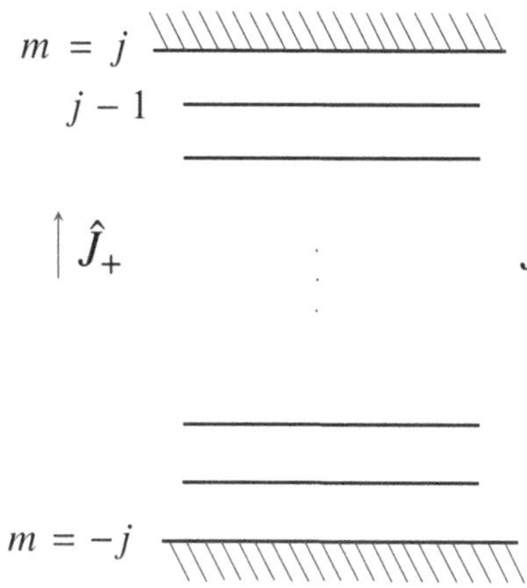

Fig. 15.5 Ladder structure of angular-momentum eigenstates $|j, m\rangle$. The raising and lowering operators $\hat{J}_+$ and $\hat{J}_-$ connect states with $m \to m \pm 1$, terminating at the extremal states $m = \pm j$ "ceiling" and "pavement" in the figure)

where the last step is justified by the fact that $\langle \beta, m|(\hat{J}_x^2 + \hat{J}_y^2)|\beta, m\rangle \geq 0$, since the operators are Hermitian; therefore $m^2 \leq \beta^2$.

There must exist a j such that $\hat{J}_+|\beta, j\rangle = 0$, hence:

$$\beta^2\hbar^2|\beta, j\rangle = \hat{J}^2|\beta, j\rangle = \left((\hat{J}_x - i\hat{J}_y)(\hat{J}_x + i\hat{J}_y) - i[\hat{J}_x, \hat{J}_y] + \hat{J}_z^2\right)|\beta, j\rangle$$
$$= \left(\hat{J}_-\hat{J}_+ + \hat{J}_z(\hat{J}_z + \hbar)\right)|\beta, j\rangle = j\hbar(j\hbar + \hbar)|\beta, j\rangle$$

Thus one finds $\beta^2 = j(j+1)$. To find the bottom state, one alternatively writes:

$$\beta^2\hbar^2|\beta, m\rangle = \hat{J}^2|\beta, m\rangle = \left(\hat{J}_+\hat{J}_- + \hat{J}_z(\hat{J}_z - \hbar)\right)|\beta, m\rangle = m\hbar(m\hbar - \hbar)|\beta, m\rangle$$

The equation $j(j+1) = m(m-1)$ has solutions $m = -j, j+1$, but the second solution is not acceptable since the top state is $m = j$; thus one finds the full constraint $-j \leq m \leq j$. The final situation is depicted in Fig. 15.5.

The problem of determining the irreducible representations of SO(3) is therefore solved: the irreducible representations are labeled by j and have dimension $2j+1$. Necessarily $2j+1 \in \mathbb{N}$, hence, at the level of the Lie algebra, j may be integer or half-integer. However, only integer j correspond to single-valued representations of SO(3) itself. Half-integer j correspond to representations of the double cover SU(2).

To find the ladder coefficients:

$$\begin{aligned}||\hat{J}_\pm|j,m\rangle||^2 &= \langle j,m|\hat{J}_\mp\hat{J}_\pm|j,m\rangle = \langle j,m|\left(\hat{J}^2 - \hat{J}_z(\hat{J}_z \pm \hbar)\right)|j,m\rangle \\ &= [j(j+1) - m(m\pm 1)]\,\hbar^2.\end{aligned}$$

Choosing conventionally the positive solution after taking the square root (known as the Condon-Shortley phase convention), one finds:

$$\hat{J}_\pm|j,m\rangle = \hbar\sqrt{(j\mp m)(j\pm m+1)}|j,m\pm 1\rangle. \tag{15.15}$$

The constraint $-j \le m \le j$ is thus manifest. Thus each irreducible representation is characterized by a highest weight j, and consists of a finite ladder of $2j+1$ states.

The matrix elements of the operators $\hat{J}_\pm$, $\hat{J}_z$ (square matrices of dimension $2j+1$) are:

$$\langle j,m'|\hat{J}_z|j,m\rangle = \hbar m\delta_{m',m} \tag{15.16}$$

$$\langle j,m'|\hat{J}_\pm|j,m\rangle = \hbar\sqrt{(j\mp m)(j\pm m+1)}\delta_{m',m\pm 1}. \tag{15.17}$$

These diagonal and near-diagonal matrices provide the irreducible representations (labelled by j) of the Lie algebra $\mathfrak{so}(3)$. For integer j, they exponentiate to single-valued representations of SO(3). For half-integer j, they define representations of its double cover SU(2).

For the representation $j=1$ one has:

$$J_z = \hbar\begin{bmatrix}-1 & 0 & 0\\ 0 & 0 & 0\\ 0 & 0 & 1\end{bmatrix} \quad J_+ = \sqrt{2}\hbar\begin{bmatrix}0 & 0 & 0\\ 1 & 0 & 0\\ 0 & 1 & 0\end{bmatrix} \quad J_- = \sqrt{2}\hbar\begin{bmatrix}0 & 1 & 0\\ 0 & 0 & 1\\ 0 & 0 & 0\end{bmatrix}$$

These matrices are analogous to those found for the vector representation V of SO(3), with the difference that they are not expressed in the Cartesian basis (x, y, z) but in the spherical one $\left(-\frac{1}{\sqrt{2}}(x+iy),\ z,\ \frac{1}{\sqrt{2}}(x-iy)\right)$.

The matrix representation of $\hat{J}_z$ depends on the chosen ordering of the basis states. If one works in the basis

$$\{|j{=}1,m\rangle\} = \{|1,-1\rangle,\ |1,0\rangle,\ |1,1\rangle\},$$

the action of the operator is

$$\hat{J}_z|1, m\rangle = \hbar m\,|1, m\rangle,$$

and therefore its matrix representation is

$$J_z = \hbar \begin{pmatrix} -1 & 0 & 0 \\ 0 & 0 & 0 \\ 0 & 0 & 1 \end{pmatrix}.$$

This choice is consistent with the corresponding matrix for $\hat{J}_+$, which raises the magnetic quantum number m by one unit,

$$\hat{J}_+|1, m\rangle \propto |1, m+1\rangle,$$

and is represented by a matrix with nonvanishing entries only on the subdiagonal.

If instead the basis is ordered as $\{|1, 1\rangle, |1, 0\rangle, |1, -1\rangle\}$, the matrix representation of $\hat{J}_z$ would be $\hbar\,\mathrm{diag}(1, 0, -1)$.

At first sight, this statement may seem to suggest a different choice of basis compared to the basis $\left(-\frac{1}{\sqrt{2}}(x+iy),\ z,\ \frac{1}{\sqrt{2}}(x-iy)\right)$ quoted above. However, there is no inconsistency.

The matrices J_z, $J_\pm$ are written in the basis of angular momentum eigenstates

$$\{|j{=}1, m\rangle\} = \{|1, -1\rangle,\ |1, 0\rangle,\ |1, 1\rangle\},$$

that is, in the basis that diagonalizes $\hat{J}_z$. This is a basis in the abstract representation space of the group, labeled by the magnetic quantum number m.

On the other hand, the sentence refers to the basis of the underlying *vector space* on which the group acts. The defining (vector) representation of SO(3) is usually expressed in the Cartesian basis (x, y, z). If one instead introduces the spherical basis

$$\left(-\frac{1}{\sqrt{2}}(x+iy),\ z,\ \frac{1}{\sqrt{2}}(x-iy)\right),$$

the generators of rotations take exactly the same matrix form as those of the $j = 1$ representation written in the $|1, m\rangle$ basis.

Thus, the matrices displayed for $j = 1$ coincide with those of the vector representation of SO(3) after a change from the Cartesian basis to the spherical basis. The difference lies not in the ordering of the states, but in the choice of basis for the vector space itself.

Having determined the irreducible representations, we now compute their characters, which depend only on the rotation angle.

15.6 Characters

The characters of the representation SO(3) can be computed by recalling that conjugacy classes depend only on the value of the rotation angle and not on the choice of the rotation axis. Therefore, it is sufficient to restrict the calculation to rotations about the z-axis. By exponentiation of Eq. (15.16), and recalling the previously derived relation between m and j, we get:

$$\mathbf{R}_3(\phi) = e^{-i\mathbf{X}_3\phi} = e^{-\frac{i}{\hbar}J_z\phi} = \mathrm{diag}\left(e^{-ij\phi}, e^{-i(j-1)\phi}, \ldots, e^{i(j-1)\phi}, e^{ij\phi}\right).$$

We notice that the trace can be computed exactly because it is given by a geometric series of constant ratio $e^{-i\varphi}$. It follows that:

$$\chi^j(\phi) = \sum_{m=-j}^{j} e^{-im\phi} = e^{-ij\phi}\frac{1-e^{i(2j+1)\phi}}{1-e^{i\phi}} = \frac{e^{-i(j+1/2)\phi} - e^{i(j+1/2)\phi}}{e^{-i\phi/2} - e^{i\phi/2}}.$$

That is,

$$\chi^j(\phi) = \frac{\sin\left((j+\frac{1}{2})\phi\right)}{\sin\left(\frac{1}{2}\phi\right)}. \tag{15.18}$$

If $j \in \mathbb{N}$, the exponential series yields (the imaginary parts cancel):

$$\chi^j(\phi) = 1 + 2\sum_{k=1}^{j} \cos k\phi. \tag{15.19}$$

If instead $j = n + \frac{1}{2}$ with $n \in \mathbb{N}$, one finds:

$$\chi^j(\phi) = 2\sum_{k=0}^{n} \cos\left((k+\tfrac{1}{2})\phi\right). \tag{15.20}$$

Since SO(3) is a compact group, it is possible to define an appropriate invariant measure $d\mu(g)$ in order to perform the transformation from a discrete sum to a continuous integral. In particular, the normalization condition $\int d\mu(g) = 1$ must hold, as well as group invariance $\int d\mu(g) = \int d\mu(g')\ \forall g, g' \in G$. To obtain the correct orthogonality relations for the characters, the following measure is used:

$$\frac{1}{|G|}\sum_{g\in G}\cdots \qquad \longrightarrow \qquad \int_0^{2\pi}\cdots\frac{d\phi}{2\pi}\left(1-\cos\phi\right).$$

This weight arises from the Haar measure of SO(3) when expressed in terms of the rotation angle alone, after integrating over the possible directions of the rotation axis. Since characters depend only on the rotation angle, the full Haar measure on SO(3) reduces to an integral over the rotation angle with weight proportional to $1-\cos\phi = 2\sin^2(\phi/2)$. Indeed, this can be easily verified to be the correct integral measure that guarantees orthonormality in the vector space of characters:

$$\langle \chi^j, \chi^\ell \rangle = \frac{1}{2\pi}\int_0^{2\pi} d\phi\, 2\sin^2\left(\tfrac{1}{2}\phi\right) \frac{\sin\left((j+\frac{1}{2})\phi\right)}{\sin\frac{1}{2}\phi} \frac{\sin\left((\ell+\frac{1}{2})\phi\right)}{\sin\frac{1}{2}\phi}$$
$$= \frac{1}{2\pi}\int_0^{2\pi} d\phi\, [\cos((j-\ell)\phi) - \cos((j+\ell+1)\phi)] = \delta_{j\ell}.$$

Example: SO(3) **Rotation About** z **Acting on Spherical Harmonics**
Consider the unitary rotation operator about the z-axis by an angle α:

$$\hat{U}(R_z(\alpha)) = e^{-\frac{i}{\hbar}\alpha \hat{J}_z}.$$

The spherical harmonics $Y_{\ell m}(\theta, \phi)$ are eigenfunctions of $\hat{J}_z$:

$$\hat{J}_z Y_{\ell m} = \hbar m\, Y_{\ell m}.$$

Therefore,

$$\hat{U}(R_z(\alpha))\, Y_{\ell m} = e^{-\frac{i}{\hbar}\alpha \hat{J}_z} Y_{\ell m} = e^{-im\alpha}\, Y_{\ell m}.$$

Equivalently, using the explicit ϕ-dependence $Y_{\ell m}(\theta, \phi) \propto e^{im\phi}$, a rotation about z acts as $\phi \mapsto \phi - \alpha$, hence

$$Y_{\ell m}(\theta, \phi - \alpha) = e^{-im\alpha} Y_{\ell m}(\theta, \phi),$$

in agreement with the operator result.

Let us now consider a concrete case: $\ell = 1$. For $\ell = 1$, the basis $\{Y_{1,-1}, Y_{1,0}, Y_{1,1}\}$ transforms diagonally under $R_z(\alpha)$:

$$\hat{U}(R_z(\alpha))\Big|_{\ell=1} = \mathrm{diag}\left(e^{i\alpha},\, 1,\, e^{-i\alpha}\right),$$

(where the ordering is $m = -1, 0, 1$). The corresponding character is the trace:

$$\chi^{\ell=1}(\alpha) = e^{i\alpha} + 1 + e^{-i\alpha} = 1 + 2\cos\alpha.$$

More generally, for the $(2\ell + 1)$-dimensional irrep,

$$\chi^{\ell}(\alpha) = \sum_{m=-\ell}^{\ell} e^{-im\alpha} = \frac{\sin\left((\ell + \frac{1}{2})\alpha\right)}{\sin(\alpha/2)} = 1 + 2\sum_{k=1}^{\ell} \cos(k\alpha).$$

Reference

1. K.F. Riley, M.P. Hobson, S.J. Bence, *Mathematical Methods for Physics and Engineering*, 3rd edn. (Cambridge University Press, Cambridge, 2006)

Chapter 16
Addition of Angular Momenta in Quantum Mechanics

Abstract Having developed the representation theory of SO(3), we now turn to one of its most important physical applications: the addition of angular momenta in quantum mechanics. This formalism allows the exact treatment of composite quantum systems and plays a central role in atomic, nuclear, particle, and condensed-matter physics. This is a topic of great importance in nuclear and particle physics, but also in solid state physics where spin-orbit coupling plays a prominent role for phenomena like topological insulators, Majorana fermions, spintronics, heavy-element materials, and magnetic effects. We shall again follow the notation of Jones [1]. Excellent reference textbooks for the physics of angular momentum in quantum mechanics, besides Landau and Lifshitz [2], are Sakurai and Napolitano [3] and Cohen-Tannoudji et al. [4].

16.1 Interacting vs Non-interacting Quantum Particles

The tensor-product structure of quantum mechanics implies that the Hilbert space of a composite system is the tensor product of the Hilbert spaces of its constituents. The problem of angular-momentum addition is therefore fundamentally a problem in the decomposition of tensor-product representations.

Let us consider a quantum system of two particles. If the two particles do not interact, the eigenstates of the Hamiltonian can be written as the direct product of the two distinct eigenstates of particle 1 and of particle 2:

$$|j_1, m_1\rangle \otimes |j_2, m_2\rangle$$

If the two particles do not interact, the Hilbert space is the tensor product $\mathcal{H}_{j_1} \otimes \mathcal{H}_{j_2}$ and the rotation group acts through the tensor-product representation $D^{j_1} \otimes D^{j_2}$. This representation is in general reducible and decomposes into irreducible components, as we shall show below.

A. Zaccone, *Differential Geometry and Group Theory*,
https://doi.org/10.1007/978-3-032-27518-9_16

If an interaction term is introduced in the Hamiltonian (for instance a spin-orbit coupling $\mathcal{H}_i \propto \hat{\mathbf{L}} \cdot \hat{\mathbf{S}}$), the uncoupled basis $|j_1, m_1\rangle \otimes |j_2, m_2\rangle$ no longer diagonalizes the Hamiltonian. In this situation it becomes natural to work in the basis of eigenstates of the total angular momentum operators $\hat{\mathbf{J}}^2$ and $\hat{J}_z$.

Indeed, the Hamiltonian with the interaction term will no longer be invariant under independent rotations, but only under simultaneous rotations of the two particles which join to form a sort of "quantum rigid-body". It is therefore necessary to rewrite the direct product $|j_1, m_1\rangle \otimes |j_2, m_2\rangle$ as a linear combination of eigenstates of components of the total angular momentum, $\hat{J}^2$ and $\hat{J}_z$, with $\hat{\mathbf{J}} \equiv \hat{\mathbf{J}}_1 + \hat{\mathbf{J}}_2$ so as to respect the reduced invariance. More precisely, in the independent picture, $\hat{\mathbf{J}} = \hat{\mathbf{J}}_1 \otimes \hat{\mathbb{I}}_2 + \hat{\mathbb{I}}_1 \otimes \hat{\mathbf{J}}_2$, where $\hat{\mathbb{I}}_1$ and $\hat{\mathbb{I}}_2$ are the identity operators for particle 1 and for particle 2, respectively. One therefore has two distinct bases: the uncoupled basis

$$|j_1, m_1; j_2, m_2\rangle \equiv |j_1, m_1\rangle \otimes |j_2, m_2\rangle$$

and the coupled one

$$|j_1, j_2; j, m\rangle.$$

Although the presence of an interaction term couples the two angular momenta, the quantum numbers j_1 and j_2 remain good quantum numbers. This follows from the fact that typical rotationally invariant interactions, such as spin-orbit or spin-spin couplings proportional to $\hat{\mathbf{J}}_1 \cdot \hat{\mathbf{J}}_2$, commute with the operators $\hat{\mathbf{J}}_1^2$ and $\hat{\mathbf{J}}_2^2$. Consequently, the magnitudes of the individual angular momenta are conserved and their eigenvalues $j_1(j_1 + 1)\hbar^2$ and $j_2(j_2 + 1)\hbar^2$ label invariant subspaces of the Hilbert space. By contrast, the projection operators $\hat{J}_{1z}$ and $\hat{J}_{2z}$ generally do not commute with rotationally invariant interaction terms, and therefore their eigenvalues are not conserved. Rotational invariance implies instead that the total angular momentum operators $\hat{\mathbf{J}}^2$ and $\hat{J}_z$, with $\hat{\mathbf{J}} = \hat{\mathbf{J}}_1 + \hat{\mathbf{J}}_2$, commute with the full Hamiltonian. As a result, in the interacting theory the natural set of good quantum numbers is (j_1, j_2, j, m), and the coupled basis $|j_1, j_2; j, m\rangle$ diagonalizes the Hamiltonian.

Consider the uncoupled basis states

$$|j_1, m_1; j_2, m_2\rangle,$$

which are simultaneous eigenstates of the operators $\hat{J}_1^2, \hat{J}_{1z}, \hat{J}_2^2, \hat{J}_{2z}$. The state with maximal magnetic quantum numbers,

$$|j_1, j_1; j_2, j_2\rangle,$$

satisfies

$$\hat{J}_z|j_1, j_1; j_2, j_2\rangle = (j_1 + j_2)\hbar|j_1, j_1; j_2, j_2\rangle,$$

and is annihilated by both individual raising operators,

$$\hat{J}_{1+}|j_1, j_1; j_2, j_2\rangle = 0, \qquad \hat{J}_{2+}|j_1, j_1; j_2, j_2\rangle = 0.$$

Since the total raising operator is given by

$$\hat{J}_+ = \hat{J}_{1+} \otimes \mathbb{I}_2 + \mathbb{I}_1 \otimes \hat{J}_{2+},$$

it follows immediately that

$$\hat{J}_+|j_1, j_1; j_2, j_2\rangle = 0.$$

Therefore this state is a highest-weight state of the total angular momentum, and one uniquely identifies

$$|j_1, j_1; j_2, j_2\rangle = |j_1, j_2; j_1 + j_2, j_1 + j_2\rangle.$$

Starting from this highest-weight (top) state, the remaining states in the multiplet with $j = j_1 + j_2$ are obtained by repeated application of the lowering operator $\hat{J}_-$. Acting once gives

$$\hat{J}_-|j_1, j_2; j_1 + j_2, j_1 + j_2\rangle = \left(\hat{J}_{1-} + \hat{J}_{2-}\right)|j_1, j_1; j_2, j_2\rangle.$$

Using the action of the individual lowering operators,

$$\hat{J}_{1-}|j_1, j_1\rangle \propto |j_1, j_1 - 1\rangle, \qquad \hat{J}_{2-}|j_2, j_2\rangle \propto |j_2, j_2 - 1\rangle,$$

one finds that the state with magnetic quantum number $m = j_1 + j_2 - 1$ is a linear combination of uncoupled states,

$$|j_1, j_2; j_1 + j_2, j_1 + j_2 - 1\rangle = a\,|j_1, j_1 - 1; j_2, j_2\rangle + b\,|j_1, j_1; j_2, j_2 - 1\rangle,$$

where the coefficients a and b are fixed by normalization and phase conventions. Repeated application of $\hat{J}_-$ generates all states in the multiplet with fixed $j = j_1 + j_2$, producing states with $m = j_1 + j_2, j_1 + j_2 - 1, \ldots, -(j_1 + j_2)$.

Once the entire $j = j_1 + j_2$ multiplet has been constructed, one can identify a new highest-weight state orthogonal to it, corresponding to total angular momentum $j = j_1 + j_2 - 1$, and repeat the same procedure. Iterating this construction yields all allowed values of the total angular momentum in the range

$$|j_1 - j_2| \le j \le j_1 + j_2.$$

We have thus recovered, from purely algebraic considerations, the familiar triangle inequality for the addition of angular momenta. For each value of j there are $2j + 1$

states labeled by $m = -j, \ldots, j$, and therefore the total number of states is:

$$\sum_{j=|j_1-j_2|}^{j_1+j_2} (2j+1) = \sum_{j=|j_1-j_2|}^{j_1+j_2} \left((j+1)^2 - j^2\right) = (j_1+j_2+1)^2 - (j_1-j_2)^2 \tag{16.1}$$

$$= (2j_1+1)(2j_2+1) \tag{16.2}$$

where the final result is due to the fact that only the first and the last term survive in the sum, while the other terms cancel each other out.

This number $(2j_1+1)(2j_2+1)$ is the total number of states both in the uncoupled and in the coupled representations. This can be checked as follows: The total number of states of a system obtained by coupling two angular momenta with quantum numbers j_1 and j_2 is the same in both the uncoupled and the coupled representations. In the uncoupled basis, the states $|j_1, m_1; j_2, m_2\rangle$ are labeled by the magnetic quantum numbers $m_1 = -j_1, \ldots, j_1$ and $m_2 = -j_2, \ldots, j_2$, yielding a total of $(2j_1+1)(2j_2+1)$ independent states. In the coupled basis, the states $|j_1, j_2; j, m\rangle$ are labeled by the total angular momentum quantum number j, which takes values in the range $|j_1 - j_2| \le j \le j_1 + j_2$, and by its projection $m = -j, \ldots, j$. The total number of states is then given by Eq. (16.2) above and coincides with the dimension of the tensor-product Hilbert space $\mathcal{H}_{j_1} \otimes \mathcal{H}_{j_2}$. The coupled and uncoupled representations therefore provide two different orthonormal bases of the same Hilbert space, related by a unitary transformation, the coefficients of which are known as the Clebsch–Gordan coefficients (see below).

The two representations are, in effect, equivalent, with the difference that in the coupled representation m_1 and m_2 are not known, since $\hat{J}_{1,z}$ and $\hat{J}_{2,z}$ do not commute with $\hat{J}^2$, while in the uncoupled one j is not known for the same reason.

The result obtained above can now be reformulated in the language of representation theory.

16.2 Clebsch–Gordan Series

The tensor-product representation is reducible and decomposes into a direct sum of irreducible representations. Since the addition of two angular momenta with eigenvalues j_1, j_2 results in an angular momentum with eigenvalue $|j_1 - j_2| \le j \le j_1 + j_2$, for the associated irreducible representations one has:

$$D^{j_1} \otimes D^{j_2} = \bigoplus_{j=|j_1-j_2|}^{j_1+j_2} D^j .$$

In particular, each irreducible representation D^j acts on a vector space of dimension $2j+1$, on which two bases are defined: the uncoupled one and the coupled one. Consequently, $D^{j_1} \otimes D^{j_2}(g)$ will be a block-diagonal matrix, where each block $D^j(g)$ on the diagonal has increasing dimension as j increases. Note that in orbital angular momentum problems these states can be realized by spherical harmonics $Y_{\ell m}(\theta, \phi)$.

Proposition 16.1 *One has $\alpha_j = 1$ for all $j \in [|j_1 - j_2|, j_1 + j_2]$, that is:*

$$D^{j_1} \otimes D^{j_2} = \bigoplus_{j=|j_1-j_2|}^{j_1+j_2} D^j \tag{16.3}$$

Proof From Eq. (15.18), assuming without loss of generality $j_1 \geq j_2$:

$$\begin{aligned}
\chi_{j_1}(\phi)\chi_{j_2}(\phi) &= \frac{e^{i(j_1+1/2)\phi} - e^{-i(j_1+1/2)\phi}}{2i \sin \frac{1}{2}\phi} \sum_{m=-j_2}^{j_2} e^{im\phi} \\
&= \frac{1}{2i \sin \frac{1}{2}\phi} \sum_{m=-j_2}^{j_2} \left(e^{i(j_1+m+1/2)\phi} - e^{-i(j_1-m+1/2)\phi}\right) \\
&= \frac{1}{2i \sin \frac{1}{2}\phi} \sum_{j=j_1-j_2}^{j_1+j_2} \left(e^{i(j+1/2)\phi} - e^{-i(j+1/2)\phi}\right) \\
&= \sum_{j=j_1-j_2}^{j_1+j_2} \frac{\sin\left((j+\frac{1}{2})\phi\right)}{\sin \frac{1}{2}\phi} = \sum_{j=j_1-j_2}^{j_1+j_2} \chi_j(\phi).
\end{aligned}$$

□

16.3 Clebsch–Gordan Coefficients

The coupled and the uncoupled representations are connected via a unitary transformation (practically, a change of basis), the coefficients of which are known as the Clebsch-Gordan coefficients. We shall compute them later explicitly on a concrete example.

While the tensor product representation $D^{j_1} \otimes D^{j_2}(g)$ acts on the uncoupled basis, each block $D^j(g)$ acts on the coupled basis. The transformation from one basis to the other is given by the *Clebsch–Gordan coefficients*:

$$|j_1, m_1; j_2, m_2\rangle = \sum_{j=|j_1-j_2|}^{j_1+j_2} \sum_{m=-j}^{j} C(j_1, j_2, j; m_1, m_2, m)|j_1, j_2; j, m\rangle. \tag{16.4}$$

Note that all these eigenvectors are orthogonal to one another, since they correspond to distinct eigenvalues of Hermitian operators:

$$\langle j_1, m_1'; j_2, m_2' \mid j_1, m_1; j_2, m_2\rangle = \delta_{m_1,m_1'}\delta_{m_2,m_2'} \tag{16.5}$$

$$\langle j_1, j_2; j', m' \mid j_1, j_2; j, m\rangle = \delta_{j,j'}\delta_{m,m'} \tag{16.6}$$

Consequently, the coefficients can be determined as:

$$C(j_1, j_2, j; m_1, m_2, m) = \langle j_1, j_2; j, m \mid j_1, m_1; j_2, m_2\rangle. \tag{16.7}$$

They can be viewed as the elements of a unitary matrix whose first index is (j, m) (new basis) and whose second is (m_1, m_2) (old basis). Moreover, these coefficients can be chosen real (and thus the matrix orthogonal) up to an arbitrary phase. From unitarity:

$$\sum_{j=|j_1-j_2|}^{j_1+j_2} \sum_{m=-j}^{j} C(j_1, j_2, j; m_1, m_2, m)\overline{C(j_1, j_2, j; m_1', m_2', m)} = \delta_{m_1,m_1'}\delta_{m_2,m_2'} \tag{16.8}$$

$$\sum_{m_1=-j_1}^{j_1} \sum_{m_2=-j_2}^{j_2} C(j_1, j_2, j; m_1, m_2, m)\overline{C(j_1, j_2, j'; m_1, m_2, m')} = \delta_{j,j'}\delta_{m,m'}. \tag{16.9}$$

The first expresses the orthogonality of the uncoupled basis and the completeness of the coupled one, while the second expresses the converse.

A more commonly used relation than Eq. (16.4) is the following:

$$|j_1, j_2; j, m\rangle = \sum_{m_1=-j_1}^{j_1} \sum_{m_2=-j_2}^{j_2} \overline{C(j_1, j_2, j; m_1, m_2, m)}|j_1, m_1; j_2, m_2\rangle. \tag{16.10}$$

In general, to determine the eigenvectors in the coupled basis, one first determines $|j_1, j_2; j, j\rangle$ and then obtains the states with $-j \le m \le j$ by repeatedly applying $\hat{J}_-$.

Through the Clebsch–Gordan coefficients it is also possible to make explicit the representation of the rotation matrices of the system starting from those of the individual angular momenta. Recalling Eq. (15.4), one has:

$$\begin{aligned}[D^j(\mathbf{R})]_{m',m} &= \langle j_1, j_2; j, m' \mid \hat{U}_{\mathbf{R}} \mid j_1, j_2; j, m\rangle \\ &= \sum_{m_1,m_1'=-j_1}^{j_1} \sum_{m_2,m_2'=-j_2}^{j_2} C(j_1, j_2, j; m_1', m_2', m')\overline{C(j_1, j_2, j; m_1, m_2, m)} \\ &\quad \times \langle j_1, m_1'; j_2, m_2' \mid \hat{U}_{\mathbf{R}} \mid j_1, m_1; j_2, m_2\rangle.\end{aligned}$$

In the uncoupled basis $\hat{U}_{\mathbf{R}}$ acts separately on each angular momentum:

$$[D^j(\mathbf{R})]_{m',m} = \sum_{m_1,m_1'=-j_1}^{j_1} \sum_{m_2,m_2'=-j_2}^{j_2} C(j_1, j_2, j; m_1', m_2', m')\overline{C(j_1, j_2, j; m_1, m_2, m)} \times [D^{j_1}(\mathbf{R})]_{m_1',m_1} [D^{j_2}(\mathbf{R})]_{m_2',m_2}. \tag{16.11}$$

This is a central formula which allows one to compute the rotation matrix for composite quantum particles starting from the independent rotations of the constituent particles. We shall use it in a worked-out example at the end of this chapter.

To make these constructions explicit for arbitrary rotations, it is convenient to introduce a standard parametrization of rotations in terms of Euler angles.

16.4 Euler Angles

From classical mechanics it is known that a generic rotation in $\mathbb{R}^3$ can be parametrized by three angles ϕ, θ, ψ, called *Euler angles*, in the following way: the first rotation by ϕ is about the z axis, the subsequent one by θ is about the new (rotated) y axis, and the last one by ψ is about the new (twice-rotated) z axis. In this way:

$$\mathbf{R}(\phi, \theta, \psi) = \mathbf{R}_3(\psi)\mathbf{R}_2(\theta)\mathbf{R}_3(\phi). \tag{16.12}$$

In the quantum framework, as seen in the previous chapter when we derived,

$$\langle j, m' \mid \hat{J}_z \mid j, m\rangle = \hbar\, m\, \delta_{m',m}.$$

rotations about the z axis have a trivial representation, since $\mathbf{R}_3(\alpha) = \exp(-i\mathbf{X}_3\alpha)$ with $\mathbf{X}_3$ diagonal; the only rotation with a non-trivial representation in Eq. (16.12) is $\mathbf{R}_2(\theta)$, for which one defines the important Wigner d-matrix $\mathbf{d}(\theta) \equiv \exp(-i\mathbf{X}_2\theta)$, with $\mathbf{X}_2$ non-diagonal, so that:

$$[D^j(\phi, \theta, \psi)]_{m',m} = e^{-im'\psi}\, [\mathbf{d}_j(\theta)]_{m',m}\, e^{-im\phi} \tag{16.13}$$

where $[D^j(\phi, \theta, \psi)]_{m',m}$ is, instead, known as the Wigner D-matrix.

We now illustrate the general formalism with the simplest nontrivial example: the coupling of two spin-$\frac{1}{2}$ particles.

16.5 Clebsch-Gordan Coefficients of Two Spin-$\frac{1}{2}$ Fermions

According to quantum statistics, fermions are particles with half-integer spin; consequently, the possible values of the total angular momentum j are only 0 and 1:

$$D^{1/2} \otimes D^{1/2} = D^1 \oplus D^0 .$$

For simplicity, the indices $j_1 = j_2 = \frac{1}{2}$ are suppressed; therefore, the uncoupled basis is denoted by $|m_1, m_2\rangle$, while the coupled basis is denoted by $|j, m\rangle$. Since the only way to obtain $m = 1$ is to have $m_1 = m_2 = \frac{1}{2}$, one immediately identifies the top state (choosing the arbitrary phase $\varphi = +1$) as

$$|1, 1\rangle = \left|\tfrac{1}{2}, \tfrac{1}{2}\right\rangle .$$

Recalling Eq. (15.15), one finds:

$$\begin{aligned}\hat{J}_-|1, 1\rangle &= \sqrt{2}\,|1, 0\rangle \\ &= \left(\hat{J}_{1,-} \otimes \mathbb{I}_2 + \mathbb{I}_1 \otimes \hat{J}_{2,-}\right)\left|\tfrac{1}{2}, \tfrac{1}{2}\right\rangle = \left| -\tfrac{1}{2}, \tfrac{1}{2}\right\rangle + \left|\tfrac{1}{2}, -\tfrac{1}{2}\right\rangle .\end{aligned}$$

By iterating the procedure one finds

$$|1, -1\rangle = \left| -\tfrac{1}{2}, -\tfrac{1}{2}\right\rangle .$$

On the other hand, the state $|0, 0\rangle$ must be a linear combination of $\left| -\frac{1}{2}, \frac{1}{2}\right\rangle$ and $\left|\frac{1}{2}, -\frac{1}{2}\right\rangle$; moreover, one must have:

$$0 = \hat{J}_+|0, 0\rangle = \hat{J}_+\left(\alpha\left| -\tfrac{1}{2}, \tfrac{1}{2}\right\rangle + \beta\left|\tfrac{1}{2}, -\tfrac{1}{2}\right\rangle\right) = (\alpha + \beta)\left|\tfrac{1}{2}, \tfrac{1}{2}\right\rangle .$$

Solving this condition and imposing normalization, one finds $\alpha = -\beta = \frac{1}{\sqrt{2}}$. The Clebsch–Gordan coefficients for two spin-$\frac{1}{2}$ particles are therefore:

$$|1, 1\rangle = \left|\tfrac{1}{2}, \tfrac{1}{2}\right\rangle , \qquad |1, 0\rangle = \frac{1}{\sqrt{2}}\left(\left|\tfrac{1}{2}, -\tfrac{1}{2}\right\rangle + \left| -\tfrac{1}{2}, \tfrac{1}{2}\right\rangle\right) , \qquad |1, -1\rangle = \left| -\tfrac{1}{2}, -\tfrac{1}{2}\right\rangle ,$$

$$|0, 0\rangle = \frac{1}{\sqrt{2}}\left(\left|\tfrac{1}{2}, -\tfrac{1}{2}\right\rangle - \left| -\tfrac{1}{2}, \tfrac{1}{2}\right\rangle\right) .$$

16.6 Wigner D-matrix and Wigner d-Matrix

Strictly speaking, the addition of angular momenta is formulated in terms of representations of SU(2), whose Lie algebra coincides with that of SO(3) but which admits also half-integer representations.

In this way one obtains:

$$\mathbf{d}_1(\theta) = \begin{bmatrix} \frac{1}{2}(1+\cos\theta) & -\frac{1}{\sqrt{2}}\sin\theta & \frac{1}{2}(1-\cos\theta) \\ \frac{1}{\sqrt{2}}\sin\theta & \cos\theta & -\frac{1}{\sqrt{2}}\sin\theta \\ \frac{1}{2}(1-\cos\theta) & \frac{1}{\sqrt{2}}\sin\theta & \frac{1}{2}(1+\cos\theta) \end{bmatrix}.$$

Recall that a generic rotation about the y-axis in the Cartesian basis (x, y, z) is given by the matrix:

$$\mathbf{R}_2(\theta) = \begin{bmatrix} \cos\theta & 0 & \sin\theta \\ 0 & 1 & 0 \\ -\sin\theta & 0 & \cos\theta \end{bmatrix}.$$

Applying this rotation to the basis $\left(-\frac{1}{\sqrt{2}}(x+iy),\ z,\ \frac{1}{\sqrt{2}}(x-iy)\right)$ one finds:

$$\begin{pmatrix} -\frac{1}{\sqrt{2}}(x'+iy') \\ z' \\ \frac{1}{\sqrt{2}}(x'-iy') \end{pmatrix} = \begin{pmatrix} -\frac{1}{\sqrt{2}}(x\cos\theta+iy+z\sin\theta) \\ -x\sin\theta+z\cos\theta \\ \frac{1}{\sqrt{2}}(x\cos\theta-iy+z\sin\theta) \end{pmatrix} = \mathbf{d}_1(\theta)\begin{pmatrix} -\frac{1}{\sqrt{2}}(x+iy) \\ z \\ \frac{1}{\sqrt{2}}(x-iy) \end{pmatrix}.$$

One therefore sees that $\mathbf{d}_1(\theta)$ represents a rotation by an angle θ about the y-axis with respect to the spherical basis $\left(-\frac{1}{\sqrt{2}}(x+iy),\ z,\ \frac{1}{\sqrt{2}}(x-iy)\right)$.

The addition of angular momenta in composite systems proceeds by iterative pairwise coupling. Since angular momentum addition is associative, a system with many constituents can be treated by successively coupling individual angular momenta.

Different coupling schemes correspond to different choices of intermediate angular momenta and are related by recoupling coefficients such as Wigner 6-j and 9-j symbols (which are closely related to the Racah W-coefficients). This formalism underlies the treatment of angular momentum in atomic and nuclear structure, where intrinsic spins and orbital angular momenta are combined subject to symmetry and antisymmetry constraints. More information can be found in the classic textbook on angular momentum by Brink and Satchler [5].

References

1. H.F. Jones, *Groups, Representations and Physics*, 2nd edn. (Taylor & Francis Ltd, Bristol/New York, 1998)
2. L.D. Landau, E.M. Lifshitz, *Quantum Mechanics: Non-relativistic Theory*. Course of Theoretical Physics, vol. 3, 3rd edn. (Pergamon Press, Oxford, 1977)
3. J.J. Sakurai, J. Napolitano, *Modern Quantum Mechanics*, 2nd edn. (Addison-Wesley, San Francisco, 2011). Originally published as Modern Quantum Mechanics by J. J. Sakurai

4. C. Cohen-Tannoudji, B. Diu, F. Laloë, *Quantum Mechanics*, vol. 2, 2nd edn. (Wiley-VCH, Hoboken, 2005)
5. D.M. Brink, G.R. Satchler, *Angular Momentum*, 3rd edn. (Oxford University Press, Oxford, 1993). Oxford Library of Physics

Chapter 17
Representation Theory of the Lorentz Group SO(3,1)

Abstract The Lorentz group is the symmetry group of Minkowski spacetime and plays a central role in modern physics. In addition to spatial rotations, it includes Lorentz boosts, whose matrix structure must be understood in order to develop a coherent representation theory. After reviewing the essentials of special relativity and introducing four-vector notation, we derive the Lie algebra of SO(3, 1), study its structure and topology, and classify its finite-dimensional representations, including Weyl and Dirac spinors. For readers who wish to refresh their knowledge of special relativity, see the book by Zangwill [1]. For a detailed treatment of the finite-dimensional representations of the Lorentz algebra the interested reader is referred to Zee [2].

17.1 Recap of Special Relativity

Let us begin by considering two reference frames S and S', where the latter moves with constant velocity $\mathbf{v}$ with respect to the former. In the framework of Newtonian mechanics, the transformations between the two frames are given by the *Galilean transformations*:

$$\begin{cases} t' = t \\ \mathbf{x}' = \mathbf{x} - \mathbf{v}\,t. \end{cases} \tag{17.1}$$

These transformations preserve the distance between two points in Euclidean 3D space:

$$\left(\mathbf{x}_1' - \mathbf{x}_2'\right)^2 = (\mathbf{x}_1 - \mathbf{x}_2)^2 . \tag{17.2}$$

The corresponding transformation for the velocity is:

$$\mathbf{u}' = \mathbf{u} - \mathbf{v}. \tag{17.3}$$

A. Zaccone, *Differential Geometry and Group Theory*,
https://doi.org/10.1007/978-3-032-27518-9_17

This transformation, however, is in contradiction with the empirical observation that the speed of light is constant in all inertial reference frames. By imposing this constraint, Einstein, in 1905, derived the *Lorentz transformations*:

$$\begin{cases} t' = \gamma \left(t - \frac{1}{c^2} \mathbf{x} \cdot \mathbf{v} \right) \\ \mathbf{x}'_\parallel = \gamma \left(\mathbf{x}_\parallel - \mathbf{v}\, t \right) \\ \mathbf{x}'_\perp = \mathbf{x}_\perp \end{cases} \tag{17.4}$$

where the parallel and perpendicular directions are defined with respect to velocity of relative motion between the two frames: $\mathbf{v}$, and

$$\gamma \equiv \left(1 - \mathbf{v}^2/c^2 \right)^{-1/2} \in [1, \infty).$$

Proposition 17.1 *Lorentz transformations preserve the speed of light: if a particle moves with velocity $|\mathbf{u}| = c$ in one inertial frame, then $|\mathbf{u}'| = c$ in any other inertial frame.*

Proof Using the chain rule:

$$\begin{aligned} \mathbf{u}'^2 = \mathbf{u}_\parallel'^2 + \mathbf{u}_\perp'^2 &= \left(\frac{d\mathbf{x}'_\parallel}{dt} \frac{dt}{dt'} \right)^2 + \left(\frac{d\mathbf{x}'_\perp}{dt} \frac{dt}{dt'} \right)^2 \\ &= \left(\frac{\gamma \left(\mathbf{u}_\parallel - \mathbf{v} \right)}{\gamma \left(1 - \mathbf{u} \cdot \mathbf{v}/c^2 \right)} \right)^2 + \left(\frac{\mathbf{u}_\perp}{\gamma \left(1 - \mathbf{u} \cdot \mathbf{v}/c^2 \right)} \right)^2 \\ &= c^2 - c^2 \left(1 - \frac{\mathbf{v}^2}{c^2} \right) \left(1 - \frac{\mathbf{u}^2}{c^2} \right) \left(1 - \frac{\mathbf{u} \cdot \mathbf{v}}{c^2} \right)^{-1} . \end{aligned}$$

Therefore one has $\|\mathbf{u}'\| \leq c$ and $\|\mathbf{u}'\| = c \iff \|\mathbf{u}\| = c$. □

For Lorentz transformations, the invariant quantity is no longer the Cartesian distance between two points in 3D Euclidean space, but the separation between two events in Minkowski spacetime:

$$c^2 \left(t_1' - t_2' \right)^2 - \left(\mathbf{x}_1' - \mathbf{x}_2' \right)^2 = c^2 \left(t_1 - t_2 \right)^2 - \left(\mathbf{x}_1 - \mathbf{x}_2 \right)^2 . \tag{17.5}$$

Also,

$$c^2 (dt)^2 - (dx)^2 - (dy)^2 - (dz)^2 = \text{const} = ds^2,$$

which vividly renders the physical intuition of a spherical light signal that propagates at constant speed c in all Cartesian directions. This is nothing but the statement that the propagation of a physical light wave is the same in all reference frames.

Using the Lorentz transformation, one indeed obtains:

$$\begin{aligned} c^2(t')^2 - (x')^2 &= \gamma^2\left[c^2\left(t - \frac{vx}{c^2}\right)^2 - (x - vt)^2\right] - y^2 - z^2 \\ &= c^2t^2 - x^2. \end{aligned}$$

To express these transformations in a form suitable for group-theoretical analysis, it is convenient to introduce a covariant four-vector notation.

17.2 Four-Vector Notation

A rotation with respect to a classical inertial frame leaves invariant the quadratic norm of a vector $\mathbf{x}$, or equivalently the scalar product of two vectors:

$$\mathbf{x} \cdot \mathbf{y} = \text{const.}$$

As a consequence, the spatial interval remains invariant,

$$ds^2 = dx^2 + dy^2 + dz^2.$$

By contrast, Lorentz transformations leave invariant not the spatial interval, but the spacetime interval

$$c^2 d\tau^2 \equiv c^2(dt)^2 - (d\mathbf{x})^2,$$

and also the combination

$$c^2 t_1 t_2 - \mathbf{x}_1 \cdot \mathbf{x}_2,$$

which can be interpreted as a scalar product if one treats ct as an additional coordinate x_0. One therefore defines the four-vector

$$x^\mu \equiv (x_0, \mathbf{x}) = (ct, \mathbf{x}), \qquad \mu = 0, 1, 2, 3.$$

One also defines

$$x_\mu \equiv (x_0, -\mathbf{x}) = (ct, -\mathbf{x}),$$

introducing the Minkowski scalar product. In this notation one has

$$x_\mu x^\mu = x^2 = x_0^2 - \mathbf{x}^2 = c^2t^2 - \mathbf{x}^2 = c^2\tau^2.$$

The scalar product of the four-vectors x_1 and x_2 is defined as

$$x_1 \cdot x_2 \equiv (x_1)_\mu (x_2)^\mu = c^2 t_1 t_2 - \mathbf{x}_1 \cdot \mathbf{x}_2.$$

A generic four-vector a^μ transforms like the position four-vector x^μ. Other four-vectors arise naturally in relativistic dynamics, for example by differentiating the position four-vector with respect to Lorentz-invariant quantities.

One therefore defines a *four-velocity* by differentiating the position four-vector x^μ not with respect to the coordinate time t (which is not invariant), but with respect to the *proper time* τ, which is invariant. The proper time is the time measured in the reference frame that is comoving with the object, that is, the reference frame in which the object is at rest. In the laboratory frame, which is in motion with respect to the object, one has:

$$d\tau = \frac{dt}{\gamma}.$$

The quantity that is physically meaningful is therefore the proper time. Thus one defines:

$$U^\mu \equiv \frac{dx^\mu}{d\tau} = \frac{dx^\mu}{dt}\frac{dt}{d\tau} = (c, \mathbf{u})\,\frac{dt}{d\tau},$$

where

$$\frac{d\tau}{dt} = \sqrt{1 - \frac{\mathbf{u}^2}{c^2}} = \frac{1}{\gamma}.$$

Analogously, the *four-momentum* is defined as:

$$P^\mu = mU^\mu = \gamma(\mathbf{u})\,(mc,\ m\mathbf{u}),$$

which is the relativistic generalization of the classical momentum.

One can decompose the four-momentum as

$$\mathbf{P} = \gamma(\mathbf{u})\,m\mathbf{u}, \qquad \frac{E}{c} = \gamma(\mathbf{u})\,mc.$$

Since the four-momentum is a four-vector, the norm of P^μ must be Lorentz invariant. Indeed, using

$$U^\mu U_\mu = c^2,$$

one finds the associated invariant:

$$P_\mu P^\mu = m^2 c^2.$$

Explicitly, this gives

$$P_\mu P^\mu = \left(\frac{E}{c}\right)^2 - \mathbf{P}^2 = m^2c^2,$$

or equivalently,

$$E^2 = m^2c^4 + \mathbf{P}^2c^2.$$

Solving for the energy, one obtains

$$E = \sqrt{m^2c^4 + \mathbf{P}^2c^2},$$

which is the relativistic energy-momentum relation (widely used for relativistic kinematic calculations of collisions etc.), replacing the classical kinetic-energy expression $T = \mathbf{P}^2/(2m)$. In particular, for a particle at rest ($\mathbf{P} = 0$) one finds

$$E = mc^2.$$

In the non-relativistic limit $|\mathbf{P}| \ll mc$, one may expand the square root:

$$E = mc^2\sqrt{1 + \frac{\mathbf{P}^2}{m^2c^2}} \simeq mc^2 + \frac{\mathbf{P}^2}{2m},$$

where the approximation

$$(1 + x)^{1/2} \simeq 1 + \frac{x}{2}$$

has been used. Thus, in this limit the kinetic energy is

$$T = E - mc^2 \simeq \frac{\mathbf{P}^2}{2m},$$

recovering the classical result.

Finally, in the non-relativistic regime the relation

$$\mathbf{P} = \frac{E}{c^2}\,\mathbf{u}$$

reduces to

$$\mathbf{P} \simeq m\mathbf{u},$$

as expected.

17.3 Finding a Matrix Representation for the Boosts

Recall that a rotation in the xy-plane about the z-axis is given by:

$$\begin{cases} x_1' = x_1 \cos\theta - x_2 \sin\theta \\ x_2' = x_1 \sin\theta + x_2 \cos\theta \\ x_3' = x_3 \\ x_0' = x_0 \end{cases}$$

To uncover the algebraic structure of Lorentz boosts, it is convenient to parametrize them in terms of a real parameter called the rapidity, ζ.

It is convenient to introduce the rapidity parameter ζ defined by

$$\gamma = \cosh\zeta, \qquad \gamma\frac{v}{c} = \sinh\zeta.$$

Since $\cosh^2\zeta - \sinh^2\zeta = 1$, this parametrization automatically satisfies the identity

$$\gamma^2 - \gamma^2\frac{v^2}{c^2} = 1.$$

Then one also has

$$\tanh\zeta = \frac{v}{c}, \qquad \sinh\zeta = \sqrt{\gamma^2 - 1} = \gamma\frac{v}{c}.$$

One obtains the following representation of a Lorentz boost:

$$\begin{cases} x_1' = x_1 \\ x_2' = x_2 \\ x_3' = x_3 \cosh\zeta - x_0 \sinh\zeta \\ x_0' = -x_3 \sinh\zeta + x_0 \cosh\zeta \end{cases} \tag{17.6}$$

(the boost direction is along the z-axis; note the minus sign in both cases).

This is necessary in order to make the hyperbolic norm $x_0^2 - x^2$ invariant, as a consequence of $\cosh^2\zeta - \sinh^2\zeta = 1$.

Clearly, despite the formal similarity, sinh and cosh are profoundly different from sin and cos. Nonetheless, this trick shall allow us to put the boosts also in matrix form, in order to give a coherent matrix representation to all the symmetry transformations contained in SO(3, 1).

We are now in a position to give a precise group-theoretical definition of the Lorentz group.

17.4 The Lorentz Group

We are now able to provide a working definition of the Lorentz group.

Before doing so, we should note that the Lorentz transformation Eq. (17.6) is a boost along the z-axis and is not a rotation. These transformations do not form a group by themselves, because a product of two boosts necessarily includes a rotation. More precisely, pure boosts along different spatial directions do not form a closed subgroup, because the product of two non-collinear boosts produces a boost combined with a spatial rotation (the Thomas rotation).

The Lorentz group SO(3, 1) includes all linear transformations (boosts, rotations), apart from parity and time reversal, that leave the spacetime interval

$$c^2\tau^2 = x_0^2 - \mathbf{x}^2 = \Delta s^2$$

invariant.

The proper orthochronous Lorentz group $\mathrm{SO}^+(3, 1)$ is the component connected to the identity. It excludes spatial inversions,

$$x_0 \to x_0, \qquad \mathbf{x} \to -\mathbf{x},$$

since these are not continuously connected to the identity. Among these transformations one distinguishes several subgroups. The subgroup with $\det\Lambda = 1$ is the *proper Lorentz group* SO(3, 1). The subgroup that is continuously connected to the identity and preserves the direction of time is the *proper orthochronous Lorentz group* $\mathrm{SO}^+(3, 1)$.

Treating x^μ as a column vector, Lorentz transformations are represented by a real 4×4 matrix Λ, such that

$$\mathbf{x}' = \boldsymbol{\Lambda}\mathbf{x}.$$

By construction, the invariant scalar product can be written in matrix form as

$$x^\mu x_\mu = x^{\mathrm{T}}\eta x,$$

where η is the Minkowski metric tensor. Therefore, Lorentz transformations satisfy

$$x'^{\mathrm{T}}\eta x' = x^{\mathrm{T}}\eta x,$$

which implies the defining condition

$$\Lambda^{\mathrm{T}}\eta\Lambda = \eta.$$

a result derived in Chap. 2, Sect. 2.2. This condition is called *pseudo-orthogonality* and generalizes the orthogonality condition of the group O(4) to the group O(3, 1), corresponding to three negative signs and one positive sign. Taking the determinant of Eq. (17.4), one finds

$$(\det \Lambda)^2 = 1.$$

By choosing $\det \Lambda = 1$, one obtains the group SO(3, 1), which is a subgroup of O(3, 1).

17.5 Digression: Double-Coverings

The study of spinor representations requires understanding the double-covering structure of the Lorentz group. We first recall the analogous situation for SO(3) and its relationship to SU(2). This digression will help us to better understand how the Lorentz group relates to complex-valued representations of lower dimension, which is of crucial importance in physics, for the spinorial representation of elementary particles. Recall that the three Pauli matrices generate the group of 2×2 complex unitary matrices with unit determinant, namely SU(2). These matrices are parametrized by a unit vector $\mathbf{n}$ and an angle θ. The same parametrization holds for SO(3); therefore, one can consider a homomorphism SU(2) $\rightarrow$ SO(3). This homomorphism is not one-to-one, since its kernel is nontrivial. In particular, SO(3) is 2π-periodic, whereas SU(2) is 4π-periodic. Denoting by $\mathbf{U}_\mathbf{n}(\theta)$ a generic element of SU(2), in the fundamental representation (generated by the Pauli matrices, since $[\sigma_a/2, \sigma_b/2] = i\epsilon^{abc}\sigma_c/2$) one has:

$$\mathbf{U}_\mathbf{n}(\theta) = \exp\left(-\frac{i}{2}\boldsymbol{\sigma}\cdot\mathbf{n}\,\theta\right) = \mathbb{I}\cos\frac{\theta}{2} - i\,\boldsymbol{\sigma}\cdot\mathbf{n}\,\sin\frac{\theta}{2}, \tag{17.7}$$

where $\boldsymbol{\sigma} \equiv (\sigma_1, \sigma_2, \sigma_3)$. One therefore sees that $\mathbf{U}_\mathbf{n}(2\pi) = -\mathbb{I}$, whereas $\mathbf{R}_\mathbf{n}(2\pi) = \mathbb{I}$; in general, two distinct elements of SU(2) are mapped to the same element of SO(3).

In order to restore a one-to-one mapping, consider the vector defined as:

$$X \equiv \mathbf{x}\cdot\boldsymbol{\sigma} = x\sigma_1 + y\sigma_2 + z\sigma_3 = \begin{bmatrix} z & x - iy \\ x + iy & -z \end{bmatrix}.$$

One sees that X is Hermitian and traceless, and that a generic element $\mathbf{U} \in$ SU(2) acts on it as $X' = \mathbf{U}^\dagger X \mathbf{U}$. Recalling that $\mathbf{U}$ is unitary, one has $\det X' = \det X$, and since $\det X = -\mathbf{x}^2$, $\mathbf{U}$ preserves the norm of the vector; this suggests that the transformation corresponds to a rotation $\mathbf{R} \in$ SO(3). To show this explicitly, consider $\mathbf{n} = (0, 0, 1)$:

$$\mathbf{U}_3(\theta) = \mathbb{I}\cos\frac{\theta}{2} - i\sigma_3\sin\frac{\theta}{2} = \begin{bmatrix} \cos\frac{\theta}{2} - i\sin\frac{\theta}{2} & 0 \\ 0 & \cos\frac{\theta}{2} + i\sin\frac{\theta}{2} \end{bmatrix} = \begin{bmatrix} e^{-i\frac{\theta}{2}} & 0 \\ 0 & e^{i\frac{\theta}{2}} \end{bmatrix}.$$

To determine the action on X, it suffices to determine the action on the Pauli matrices:

$$\mathbf{U}_3^{\dagger}\sigma_1\mathbf{U}_3 = \begin{bmatrix} e^{i\frac{\varphi}{2}} & 0 \\ 0 & e^{-i\frac{\varphi}{2}} \end{bmatrix} \begin{bmatrix} 0 & 1 \\ 1 & 0 \end{bmatrix} \begin{bmatrix} e^{-i\frac{\varphi}{2}} & 0 \\ 0 & e^{i\frac{\varphi}{2}} \end{bmatrix} = \begin{bmatrix} 0 & e^{i\varphi} \\ e^{-i\varphi} & 0 \end{bmatrix} = \sigma_1 \cos\varphi - \sigma_2 \sin\varphi,$$

$$\mathbf{U}_3^{\dagger}\sigma_2\mathbf{U}_3 = \begin{bmatrix} e^{i\frac{\varphi}{2}} & 0 \\ 0 & e^{-i\frac{\varphi}{2}} \end{bmatrix} \begin{bmatrix} 0 & -i \\ i & 0 \end{bmatrix} \begin{bmatrix} e^{-i\frac{\varphi}{2}} & 0 \\ 0 & e^{i\frac{\varphi}{2}} \end{bmatrix} = \begin{bmatrix} 0 & -ie^{i\varphi} \\ ie^{-i\varphi} & 0 \end{bmatrix}$$
$$= \sigma_1 \sin\varphi + \sigma_2 \cos\varphi,$$

$$\mathbf{U}_3^{\dagger}\sigma_3\mathbf{U}_3 = \begin{bmatrix} e^{i\frac{\varphi}{2}} & 0 \\ 0 & e^{-i\frac{\varphi}{2}} \end{bmatrix} \begin{bmatrix} 1 & 0 \\ 0 & -1 \end{bmatrix} \begin{bmatrix} e^{-i\frac{\varphi}{2}} & 0 \\ 0 & e^{i\frac{\varphi}{2}} \end{bmatrix} = \begin{bmatrix} 1 & 0 \\ 0 & -1 \end{bmatrix} = \sigma_3.$$

Therefore:

$$\begin{aligned} X' &= \mathbf{U}_3^{\dagger} X \mathbf{U}_3 \\ &= (\sigma_1 \cos\varphi - \sigma_2 \sin\varphi)\, x + (\sigma_1 \sin\varphi + \sigma_2 \cos\varphi)\, y + \sigma_3 z \\ &= \sigma_1(x\cos\varphi + y\sin\varphi) + \sigma_2(-x\sin\varphi + y\cos\varphi) + \sigma_3 z \equiv \sigma_1 x' + \sigma_2 y' + \sigma_3 z'. \end{aligned}$$

This is precisely a rotation by an angle φ about the z-axis, thus confirming that the action of $\mathbf{U}_{\mathbf{n}}(\theta) \in \mathrm{SU}(2)$ is analogous to the action of $\mathbf{R}_{\mathbf{n}}(\theta) \in \mathrm{SO}(3)$. Finally, noting that $-\mathbf{U}$ produces the same rotation as $\mathbf{U}$, one can state that this is exactly the homomorphism $\mathrm{SU}(2) \to \mathrm{SO}(3)$.

Since the pair $\{\mathbf{U}, -\mathbf{U}\}$ is mapped to the same rotation, topologically one says that SU(2) is a double covering of SO(3). Here, the word "covering" is to be intended with the same meaning that it has in topology, cf. Chap. 6, Sect. 6.1.3. See Fig. 17.1 for an illustration of this phenomenon.

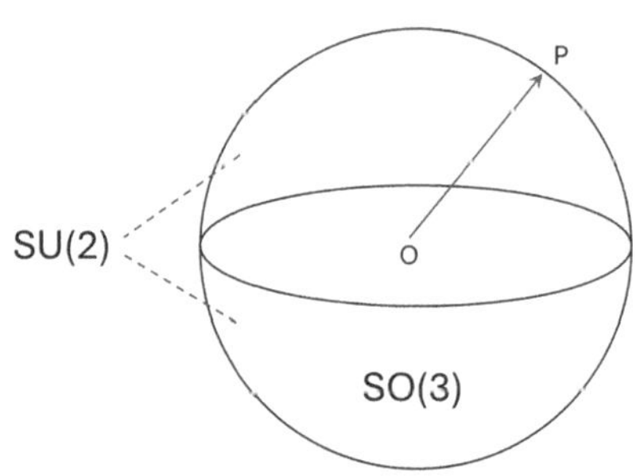

Fig. 17.1 Schematic illustration of the double covering of SO(3) by SU(2). The sphere represents the rotation group SO(3), with points corresponding to spatial rotations parametrized by an axis and an angle. The arrow from the origin O to the point P represents a finite rotation by an angle φ. The dashed connections indicate that two distinct elements of SU(2), differing by a sign, are mapped to the same element of SO(3). This reflects the fact that SU(2) is a double cover of SO(3), and that a 2π rotation acts trivially in SO(3) but nontrivially in SU(2)

The kernel of the homomorphism is therefore $\{\mathbb{I}, -\mathbb{I}\} \triangleleft \mathrm{SU}(2)$, and since $\{\mathbb{I}, -\mathbb{I}\} \cong \mathbb{Z}_2$, one has:

$$\mathrm{SO}(3) \cong \mathrm{SU}(2)/\mathbb{Z}_2. \tag{17.8}$$

In an analogous way, SO(3, 1) is related to the group SL(2, $\mathbb{C}$) of 2×2 complex matrices with unit determinant by a homomorphism that yields a double covering. Defining $\sigma_\mu \equiv (1, \boldsymbol{\sigma})$, given a four-vector v_μ one can construct an associated matrix $V = v_\mu \sigma^\mu$. Recalling that $\sigma_a \sigma_b = \mathbb{I}\delta_{ab} + i\epsilon^{abc}\sigma_c$, one can show that

$$v_\mu = \frac{1}{2}\operatorname{tr}\left(\tilde{\sigma}_\mu V\right), \qquad \tilde{\sigma}_\mu \equiv (1, -\boldsymbol{\sigma}),$$

with the same components as σ^μ except for the sign. Explicitly,

$$V = \begin{bmatrix} v_0 - v_3 & -v_1 + iv_2 \\ -v_1 - iv_2 & v_0 + v_3 \end{bmatrix},$$

and therefore $\det V = v^2$. Given $A \in \mathrm{SL}(2, \mathbb{C})$ one defines

$$V' = AVA^\dagger.$$

This transformation preserves $\det V$ and therefore the Minkowski norm $v_\mu v^\mu$. Since the pair $\{\mathbf{A}, -\mathbf{A}\}$ is mapped to the same element of SO(3, 1), one again has a double covering:

$$\mathrm{SO}(3, 1) \cong \mathrm{SL}(2, \mathbb{C})/\mathbb{Z}_2. \tag{17.9}$$

More precisely, the proper orthochronous Lorentz group $\mathrm{SO}^+(3, 1)$ is double-covered by SL(2, $\mathbb{C}$):

$$\mathrm{SO}^+(3, 1) \cong \mathrm{SL}(2, \mathbb{C})/\mathbb{Z}_2.$$

Finally, one may note that $\mathrm{SO}(3) < \mathrm{SO}(3, 1)$ and $\mathrm{SU}(2) < \mathrm{SL}(2, \mathbb{C})$.

Having identified the generators of rotations and boosts, we now analyze their algebraic structure.

17.6 Lie Generators of SO(3, 1)

Since $\mathrm{SO}(3) \subset \mathrm{SO}(3, 1)$, the generators of rotations in $\mathbb{R}^3$ are also generators of rotations in Minkowski space. In particular, rotations have no effect on the timelike coordinate; therefore, their generators are just the same as those of SO(3):

$$X_1 = \begin{bmatrix} 0 & 0 & 0 & 0 \\ 0 & 0 & 0 & 0 \\ 0 & 0 & 0 & -i \\ 0 & 0 & i & 0 \end{bmatrix} \qquad X_2 = \begin{bmatrix} 0 & 0 & 0 & 0 \\ 0 & 0 & 0 & i \\ 0 & 0 & 0 & 0 \\ 0 & -i & 0 & 0 \end{bmatrix} \qquad X_3 = \begin{bmatrix} 0 & 0 & 0 & 0 \\ 0 & 0 & -i & 0 \\ 0 & i & 0 & 0 \\ 0 & 0 & 0 & 0 \end{bmatrix}$$

As for boosts, they are represented by the boost matrices, cf. Eq. (17.6):

$$\mathbf{B}_1(\zeta) = \begin{bmatrix} \cosh\zeta & -\sinh\zeta & 0 & 0 \\ -\sinh\zeta & \cosh\zeta & 0 & 0 \\ 0 & 0 & 1 & 0 \\ 0 & 0 & 0 & 1 \end{bmatrix} \qquad \mathbf{B}_2(\zeta) = \begin{bmatrix} \cosh\zeta & 0 & -\sinh\zeta & 0 \\ 0 & 1 & 0 & 0 \\ -\sinh\zeta & 0 & \cosh\zeta & 0 \\ 0 & 0 & 0 & 1 \end{bmatrix}$$

$$\mathbf{B}_3(\zeta) = \begin{bmatrix} \cosh\zeta & 0 & 0 & -\sinh\zeta \\ 0 & 1 & 0 & 0 \\ 0 & 0 & 1 & 0 \\ -\sinh\zeta & 0 & 0 & \cosh\zeta \end{bmatrix}$$

Recalling that Lie generators are defined as

$$Y_j = i\,\frac{d}{d\zeta}\mathbf{B}_j(\zeta)\bigg|_{\zeta=0},$$

one finds:

$$Y_1 = \begin{bmatrix} 0 & -i & 0 & 0 \\ -i & 0 & 0 & 0 \\ 0 & 0 & 0 & 0 \\ 0 & 0 & 0 & 0 \end{bmatrix} \qquad Y_2 = \begin{bmatrix} 0 & 0 & -i & 0 \\ 0 & 0 & 0 & 0 \\ -i & 0 & 0 & 0 \\ 0 & 0 & 0 & 0 \end{bmatrix} \qquad Y_3 = \begin{bmatrix} 0 & 0 & 0 & -i \\ 0 & 0 & 0 & 0 \\ 0 & 0 & 0 & 0 \\ -i & 0 & 0 & 0 \end{bmatrix}$$

In compact form:

$$\left[Y_j\right]_\mu^{\ \nu} = -i\left(\eta_{\mu 0}\,\eta_j^{\ \nu} - \eta_{\mu j}\,\eta_0^{\ \nu}\right). \tag{17.10}$$

Proposition 17.2 *For the generators of* SO(3, 1) *the following commutation relations hold:*

$$\left[X_i, X_j\right] = i\epsilon^{ijk} X_k, \tag{17.11}$$

$$\left[Y_i, Y_j\right] = -i\epsilon^{ijk} X_k, \tag{17.12}$$

$$\left[X_i, Y_j\right] = i\epsilon^{ijk} Y_k. \tag{17.13}$$

Proof By direct calculation of the matrix products. □

One therefore sees that the Lie algebra $so(3,1)$ contains the subalgebra $so(3)$, generated by the rotation operators X_i. By contrast, the boost generators Y_i do not form a closed Lie subalgebra, since their commutator produces a rotation generator:

$$[K_i, K_j] = -i\epsilon_{ijk} J_k.$$

It is possible to simplify the algebra SO(3, 1) by defining new generators.

Proposition 17.3 *Defining the following generators of* SO(3, 1)*:*

$$X_j^{\pm} \equiv \frac{1}{2}\left(X_j \pm iY_j\right), \tag{17.14}$$

the following commutation relations hold:

$$\left[X_i^{\pm}, X_j^{\pm}\right] = i\epsilon^{ijk} X_k^{\pm}, \tag{17.15}$$

$$\left[X_i^{\pm}, X_j^{\mp}\right] = 0. \tag{17.16}$$

Proof By direct calculation of the matrix products. □

These six generators split into two mutually independent sets forming two closed subalgebras. Upon complexification the Lorentz algebra decomposes as

$$so(3,1)_{\mathbb{C}} \cong su(2)_{\mathbb{C}} \oplus su(2)_{\mathbb{C}}.$$

which underlies the classification of finite-dimensional representations.

Although the full Lorentz group SO(3, 1) is double-covered by SL(2, $\mathbb{C}$), its rotation subgroup SO(3) is double-covered by SU(2). Moreover, the decomposition of the Lorentz algebra into two $su(2)$ algebras ensures that representations of SU(2) naturally classify the spin content of relativistic fields, as we shall see below. The Lie algebra associated with the proper Lorentz group SO(3, 1) is denoted by $so(3,1)$ and is defined as the algebra of all generators of infinitesimal Lorentz transformations, endowed with the commutator as Lie bracket. Explicitly, elements of $so(3,1)$ are linear combinations of the generators of spatial rotations and boosts, and their commutation relations completely determine the *local* structure of the Lorentz group.

Because each set $\{X_i^+\}$ and $\{X_i^-\}$ in Eqs. (17.15)–(17.16) closes under commutation and generates an independent $su(2)$ Lie algebra, the Lorentz algebra admits the decomposition

$$so(3,1) \cong su(2) \oplus su(2).$$

This result shows that the generators of Lorentz transformations can be organized into two commuting angular-momentum algebras, each with the structure of ordinary rotations. Irreducible representations of the Lorentz group are therefore labeled by a pair of non-negative half-integers (j_1, j_2), corresponding to the spins associated with each $su(2)$ factor. The representation (j_1, j_2) has dimension $(2j_1 + 1)(2j_2+1)$, and its restriction to the rotation subgroup $SO(3) \subset SO(3, 1)$ determines the physical spin content of the relativistic field. In particular, finite-dimensional representations of the Lorentz group are necessarily non-unitary, reflecting the non-compact nature of SO(3, 1). More precisely, because SO(3, 1) is non-compact, it admits no non-trivial finite-dimensional unitary representations.

We shall study the hierarchy of relativistic spinor representations in the next section, because these play a fundamental role in elementary particle physics and in field theory. In particular, we shall classify the finite-dimensional representations of the Lorentz algebra [2].

17.7 Classification of Relativistic Spinor Representations

One typically starts from the representation (0, 0), i.e. $j_1 = 0, \; j_2 = 0$, which is the trivial one. Upon increasing the dimension of the representation, the next one is the Weyl representation, which, given its importance in physics, is discussed in detail below.

17.7.1 The Weyl Spinors and the Weyl Representation

Starting from the Lorentz algebra, we derive the generators acting on left- and right-handed two-component spinors (cf. Sect. 16.4 in Chap. 16) and show explicitly the splitting into two commuting $su(2)$ algebras.

The left-handed Weyl spinor transforms in the irreducible representation

$$\left(\tfrac{1}{2}, 0\right),$$

with generators

$$J_i = \frac{\sigma_i}{2}, \qquad K_i = +i\,\frac{\sigma_i}{2},$$

(the sign convention depends on the definition of the boost generators), so that

$$X_i = \frac{1}{2}(J_i + iK_i) = \frac{\sigma_i}{2}, \qquad Y_i = \frac{1}{2}(J_i - iK_i) = 0.$$

The right-handed Weyl spinor transforms in the irreducible representation

$$\left(0, \tfrac{1}{2}\right),$$

with generators

$$J_i = \frac{\sigma_i}{2}, \qquad K_i = -i\,\frac{\sigma_i}{2},$$

so that

$$X_i = 0, \qquad Y_i = \frac{\sigma_i}{2}.$$

Here the operators

$$X_i \equiv \frac{1}{2}(J_i + iK_i), \qquad Y_i \equiv \frac{1}{2}(J_i - iK_i)$$

generate two commuting $su(2)$ algebras. The above results can be derived or justified as follows.

Let J_i (rotations) and K_i (boosts) satisfy the Lorentz algebra

$$[J_i, J_j] = i\epsilon_{ijk}J_k, \qquad [J_i, K_j] = i\epsilon_{ijk}K_k, \qquad [K_i, K_j] = -i\epsilon_{ijk}J_k.$$

Next, define

$$X_i \equiv \frac{1}{2}(J_i + iK_i), \qquad Y_i \equiv \frac{1}{2}(J_i - iK_i).$$

One finds

$$[X_i, X_j] = i\epsilon_{ijk}X_k, \qquad [Y_i, Y_j] = i\epsilon_{ijk}Y_k, \qquad [X_i, Y_j] = 0.$$

Thus, the complexified Lorentz algebra decomposes as

$$so(3,1)_{\mathbb{C}} \cong su(2) \oplus su(2).$$

Its irreducible representations are labeled by pairs (j_L, j_R). The Weyl spinor representations correspond to

$$\left(\tfrac{1}{2}, 0\right), \qquad \left(0, \tfrac{1}{2}\right).$$

At this point we need to delve more deeply into the two-component spinor technology. Let us introduce the 2×2 matrices

$$\sigma^\mu = (\mathbf{1}, \sigma_i), \qquad \bar{\sigma}^\mu = (\mathbf{1}, -\sigma_i),$$

where σ_i are the Pauli matrices and the sign convention corresponds to the Minkowski metric is

$$\eta_{\mu\nu} = \mathrm{diag}(1, -1, -1, -1).$$

Useful identities that can be applied here are

$$\sigma^\mu \bar{\sigma}^\nu + \sigma^\nu \bar{\sigma}^\mu = 2\eta^{\mu\nu}\mathbf{1},$$

$$\bar{\sigma}^\mu \sigma^\nu + \bar{\sigma}^\nu \sigma^\mu = 2\eta^{\mu\nu}\mathbf{1}.$$

For the left-handed (undotted) spinor representation, we define

$$(\Sigma_L)^{\mu\nu} \equiv \frac{1}{4}\left(\sigma^\mu \bar{\sigma}^\nu - \sigma^\nu \bar{\sigma}^\mu\right),$$

and for the right-handed (dotted) representation

$$(\Sigma_R)^{\mu\nu} \equiv \frac{1}{4}\left(\bar{\sigma}^\mu \sigma^\nu - \bar{\sigma}^\nu \sigma^\mu\right).$$

Finite Lorentz transformations act as

$$S_L(\Lambda) = \exp\!\left(\frac{i}{2}\,\omega_{\mu\nu}(\Sigma_L)^{\mu\nu}\right), \qquad S_R(\Lambda) = \exp\!\left(\frac{i}{2}\,\omega_{\mu\nu}(\Sigma_R)^{\mu\nu}\right),$$

where $\omega_{\mu\nu}$ are antisymmetric parameters (ω_{0i} boosts, ω_{ij} rotations) defined through the Lorentz transformation matrices, $\Lambda^\mu{}_\nu = \delta^\mu{}_\nu + \omega^\mu{}_\nu$, with the antisymmetry $\omega_{\mu\nu} = -\omega_{\nu\mu}$, which follows from the Lorentz invariance.

Using $\bar{\sigma}^0 = \sigma^0 = \mathbf{1}$ and $\bar{\sigma}^i = -\sigma^i$, one finds, for the boost generators,

$$(\Sigma_L)^{0i} = \frac{1}{4}\left(\sigma^0 \bar{\sigma}^i - \sigma^i \bar{\sigma}^0\right) = -\frac{1}{2}\sigma_i,$$

and, for the rotation generators,

$$(\Sigma_L)^{ij} = -\frac{i}{2}\epsilon_{ijk}\sigma_k.$$

Identifying the Hermitian generators with

$$J_i = \frac{1}{2}\epsilon_{ijk}\, i(\Sigma_L)^{jk}, \qquad K_i = i(\Sigma_L)^{0i},$$

one obtains

$$J_i = \frac{\sigma_i}{2}, \qquad K_i = +i\,\frac{\sigma_i}{2}.$$

Therefore

$$X_i = \frac{\sigma_i}{2}, \qquad Y_i = 0,$$

showing that the left-handed Weyl spinor transforms as

$$\left(\tfrac{1}{2}, 0\right).$$

For the right-handed generators one finds

$$(\Sigma_R)^{0i} = +\frac{1}{2}\sigma_i, \qquad (\Sigma_R)^{ij} = -\frac{i}{2}\epsilon_{ijk}\sigma_k.$$

With the same identification,

$$J_i = \frac{\sigma_i}{2}, \qquad K_i = -i\,\frac{\sigma_i}{2}.$$

Thus

$$X_i = 0, \qquad Y_i = \frac{\sigma_i}{2},$$

and the right-handed Weyl spinor transforms as

$$\left(0, \tfrac{1}{2}\right).$$

As shown in the previous section, the complexified Lorentz algebra decomposes into two commuting $su(2)$ algebras generated by X_i and Y_i. Left-handed Weyl spinors transform as $\left(\frac{1}{2}, 0\right)$ with $X_i \sim \sigma_i/2$ and $Y_i = 0$, while right-handed Weyl spinors transform as $\left(0, \frac{1}{2}\right)$ with $X_i = 0$ and $Y_i \sim \sigma_i/2$. Equivalently, boosts act with opposite signs on left- and right-handed spinors, while rotations act identically.

The representation $(\frac{1}{2}, 0)$ acts on spinors u_α with $\alpha = 1, 2$.

The operators

$$X_i \equiv \frac{1}{2}(J_i + iK_i)$$

act on u and are represented by $\frac{1}{2}\sigma_i$, whereas

$$Y_i \equiv \frac{1}{2}(J_i - iK_i)$$

act trivially on u and are therefore represented by zero.

Adding and subtracting these relations, one finds

$$X_i = \frac{1}{2}\sigma_i, \qquad iY_i = \frac{1}{2}\sigma_i,$$

where the equal sign is to be understood as "represented by".

For the representation $(0, \frac{1}{2})$ acting on v, one instead obtains

$$X_i = \frac{1}{2}\sigma_i, \qquad iY_i = -\frac{1}{2}\sigma_i,$$

which differs by a sign. This sign difference reflects the action of parity,

$$Y \longrightarrow -Y,$$

and shows that the two Weyl spinors transform chirally and are exchanged under parity.

Within this framework, the dynamics of massless spin-$\frac{1}{2}$ fields is described by the Weyl equation. For a left-handed Weyl spinor u_α, the equation of motion reads

$$\sigma^{\mu}_{\alpha\dot{\beta}}\,\partial_\mu\, u^\alpha = 0,$$

while for a right-handed Weyl spinor $v^{\dot{\alpha}}$ one has

$$\bar{\sigma}^{\mu\,\dot{\alpha}\beta}\,\partial_\mu\, v_{\dot{\alpha}} = 0.$$

These equations are Lorentz covariant precisely because the spinor fields transform in the irreducible representations $(\frac{1}{2}, 0)$ and $(0, \frac{1}{2})$ of the Lorentz group, respectively. The dot on a spinor index indicates that the object transforms in the right-handed $(0, \frac{1}{2})$ representation of the Lorentz group, whereas undotted indices correspond to the left-handed $(\frac{1}{2}, 0)$ representation. The Weyl equation thus provides the natural relativistic field equation associated with chiral, parity-violating, spinor representations.

17.7.2 The Dirac Representation

To construct a parity-invariant representation one combines the two chiral representations into their direct sum

$$\left(\tfrac{1}{2},0\right)\oplus\left(0,\tfrac{1}{2}\right),$$

which is known as the Dirac representation and acts on four-component spinors (used, for example, to describe the relativistic electron). The direct sum takes into account the fact that the parity operator maps a representation (j_1, j_2) into (j_2, j_1); therefore, any representation that incorporates parity conservation must necessarily be symmetric in j_1 and j_2.

The Dirac spinor may therefore be written as

$$\Psi=\begin{pmatrix}u_\alpha\\ v^{\dot\alpha}\end{pmatrix},$$

where u_α transforms in the $(\frac{1}{2},0)$ representation and $v^{\dot\alpha}$ transforms in the $(0,\frac{1}{2})$ representation. Under parity, these two Weyl components are exchanged, so that the full Dirac representation is invariant under parity transformations.

This construction allows one to describe relativistic fermions with parity-invariant dynamics, and leads naturally to the Dirac equation, which couples the left- and right-handed components through a mass term. In this framework, the relativistic dynamics of a Dirac spinor is governed by the Dirac equation

$$\left(i\gamma^\mu\partial_\mu-m\right)\Psi=0,$$

where the γ^μ matrices provide a representation of the Clifford algebra and act on the four-component Dirac spinor Ψ. In the chiral basis, the gamma matrices explicitly couple the left-handed and right-handed Weyl components, reflecting the direct-sum structure $(\frac{1}{2},0)\oplus(0,\frac{1}{2})$ and ensuring Lorentz covariance and parity invariance of the theory. The Dirac representation thus provides the minimal Lorentz-covariant and parity-invariant framework for describing massive fermions with half-integer spin encountered in nature, such as electrons, nucleons (protons and neutrons), and quarks.

17.7.3 The Lorentz Representation

Finally, the four-dimensional representation $\left(\frac{1}{2},\frac{1}{2}\right)$ corresponds to the vector representation acting on Minkowski four-vectors.

17.8 A Note on Non-unitarity

Finite-dimensional representations of SO(3, 1) cannot be unitary because the group is non-compact. In particular, the boost generators cannot be represented by Hermitian operators in a finite-dimensional unitary representation. This is the reason why the double covering of SO(3, 1) is given by $\mathrm{SL}(2, \mathbb{C})$, whereas for SO(4) one has

$$\mathrm{SO}(4) \cong \mathrm{SU}(2) \otimes \mathrm{SU}(2)/\mathbb{Z}_2. \tag{17.17}$$

The fundamental difference is due to the signature of the metric. A further difference between SO(3, 1) and SO(4) is that, because of boosts parametrized by $\zeta \in \mathbb{R}_0^+$, SO(3, 1) is not compact.

References

1. A. Zangwill, *Modern Electrodynamics* (Cambridge University Press, Cambridge, 2013)
2. A. Zee, *Group Theory in a Nutshell for Physicists* (Princeton University Press, Princeton, 2016)

Chapter 18
Representation Theory of the Poincaré Group

Abstract We have seen that the Lorentz group is the symmetry group of flat Minkowski spacetime and describes the transformation properties of non-interacting elementary particles in empty space. However, spacetime possesses an additional fundamental symmetry not contained in the Lorentz group: translational invariance. However, spacetime possesses an additional fundamental symmetry not contained in the Lorentz group: translational invariance. In this chapter we extend the Lorentz symmetry to include translations on the same footing as rotations and boosts. The resulting structure is the Poincaré group, which is the full symmetry group of elementary particles in empty spacetime. We then develop its algebra, representations, and Casimir invariants, which ultimately characterize particles by mass and spin. We follow the notation of Jones [1]. For further details on Weyl and Dirac spinor representations and their connection with field theory, see Zee [2].

18.1 Definition

The *Poincaré group* is defined as the group of all isometries of Minkowski spacetime that preserve the spacetime interval

$$ds^2 = c^2 t^2 - \boldsymbol{x}^2.$$

As usual, the *proper Poincaré group* is defined as the subgroup of the Poincaré group obtained by excluding inversions.

The Poincaré group contains rotations, boosts, and spacetime translations. Therefore, the most general transformation in the group is

$$\tilde{x}_\mu - \Lambda_\mu^{\nu} x_\nu + \alpha_\mu, \tag{18.1}$$

A. Zaccone, *Differential Geometry and Group Theory*,
https://doi.org/10.1007/978-3-032-27518-9_18

where $\Lambda \in SO^{+}(3, 1)$ and α_{μ} is a constant four-vector. It is immediate to verify that Eq. (17.5) is satisfied.

Having defined the Poincaré group at the level of spacetime transformations, we now turn to its Lie algebra and to explicit realizations of its generators. As in the case of the Lorentz group, understanding the algebra is the key step toward constructing its representations.

18.2 Algebra and Representation

To construct the Lie algebra of the Poincaré group, we must represent its generators explicitly. Unlike Lorentz transformations, spacetime translations cannot be represented as finite matrices acting on four-vectors. The natural framework is therefore the differential representation introduced in Chap. 9, where generators act as differential operators on functions defined over Minkowski spacetime.

One therefore introduces the differential operator

$$\partial_{\mu} \equiv \left(\frac{\partial}{\partial x_0}, -\nabla\right), \tag{18.2}$$

where the minus sign is required so that ∂_{μ} transforms in the same way as x_{μ} under parity.

As a sanity check,

$$\partial_{\mu}x^2 = \partial_{\mu}(x_0^2 - \mathbf{x}^2) = 2(x_0, \mathbf{x}) = 2x_{\mu}.$$

as expected.

Consistent with the four-vector notation that we introduced in the previous chapter, we can also define the contravariant components of the differential operator:

$$\partial^{\mu} = \left(\frac{\partial}{\partial x_{\mu}}, \nabla\right)$$

which transforms in the opposite way with respect to x_{μ}.

We shall systematically use this operator in the next section to construct a consistent differential representation for rotations and boosts of the Lorentz group.

18.2.1 Generalized Angular Momentum Operator for Rotations and Boosts

Consider an infinitesimal Lorentz transformation such that

$$\Lambda_{\mu\nu} = \eta_{\mu\nu} - \varepsilon_{\mu\nu}.$$

where the tensor $\varepsilon_{\mu\nu}$ represents a small perturbation to the metric. To rewrite Eq. (17.4) in components, note that $\tilde{x}^{\mu} = \Lambda^{\mu}_{\nu} x^{\nu}$, so invariance of $c^2\tau^2$, written as $\tilde{x}^{\alpha}\eta_{\alpha\beta}\tilde{x}^{\beta} = x^{\mu}\eta_{\mu\nu}x^{\nu}$, becomes

$$\Lambda^{\alpha}_{\mu}\eta_{\alpha\beta}\Lambda^{\beta}_{\nu} = \eta_{\mu\nu}. \tag{18.3}$$

For the infinitesimal transformation one finds

$$\eta_{\mu\nu} = \Lambda_{\beta\mu}\Lambda^{\beta}_{\nu} = (\eta_{\beta\mu} - \varepsilon_{\beta\mu})(\delta^{\beta}_{\nu} - \varepsilon^{\beta}_{\nu}) = \eta_{\mu\nu} - \varepsilon_{\mu\nu} - \varepsilon_{\nu\mu} + o(\varepsilon^2),$$

which implies

$$\varepsilon_{\nu\mu} = -\varepsilon_{\mu\nu}.$$

This antisymmetry reflects the six independent generators of the Lorentz group: three rotations and three boosts. We now construct their explicit differential representation.

Since $\varepsilon_{\mu\nu}$ is antisymmetric, $\Lambda_{\mu\nu}$ depends on six independent infinitesimal parameters, corresponding to three rotations and three boosts. From Lie theory, the action of such transformations can be written as generated by operators $L_{\mu\nu}$, which are generalized angular momentum operators:

$$L_{\mu\nu} \equiv i(x_{\mu}\partial_{\nu} - x_{\nu}\partial_{\mu}). \tag{18.4}$$

By means of this generalized angular momentum operator, a generic finite symmetry transformation belonging to the Lorentz group can be obtained by exponentiation of the generator as:

$$\exp\left(-\frac{i}{2}\,\epsilon_{\rho\sigma}\,L^{\rho\sigma}\right)$$

This can be easily verified, because we have:

$$\left(1 - \frac{i}{2}\,\varepsilon_{\rho\sigma}\,L^{\rho\sigma}\right)x_{\mu} = x_{\mu} + \frac{i}{2}\,\varepsilon_{\rho\sigma}\left(x^{\rho}\partial^{\sigma} - x^{\sigma}\partial^{\rho}\right)x_{\mu}$$

$$= x_{\mu} + \varepsilon_{\rho\sigma}\,x^{\rho}\,\delta^{\sigma}{}_{\mu} = x_{\mu} - \varepsilon_{\mu\rho}\,x^{\rho}$$

where we have used the antisymmetry of $\varepsilon_{\mu\nu}$ and the identity $\partial^{\sigma}(x_{\mu}) = \delta^{\sigma}_{\mu}$.

This identity can be written in operator form as the commutation bracket:

$$[\partial^{\sigma}, x_{\mu}] = \delta^{\sigma}{}_{\mu} \tag{18.5}$$

since $x_{\mu}\,\partial^{\sigma} = 0$.

This allows one to compute the commutation relations of the differential operators $L_{\rho\sigma}$.

Armed with these results, we are now able to compute commutation brackets between generalized angular momentum operators.

The generalized angular momentum operators satisfy

$$[L_{\mu\nu}, L_{\rho\sigma}] = -i(\eta_{\mu\rho}L_{\nu\sigma} - \eta_{\mu\sigma}L_{\nu\rho} + \eta_{\nu\sigma}L_{\mu\rho} - \eta_{\nu\rho}L_{\mu\sigma}). \tag{18.6}$$

This equation can be proved by direct computation using (18.4) and the identity (18.5).

It is convenient to define the rotation and boost generators in terms of the generalized momentum operator defined in (18.4). Using Latin indices which run over 1, 2, 3 for spatial-like components and 0 for the time-like component, we have:

$$J_i \equiv \frac{1}{2}\epsilon^{ijk}L_{jk}, \qquad K_i \equiv L_{0i}. \tag{18.7}$$

In natural units ($c = \hbar = 1$) one finds

$$J_i = -i(\mathbf{x} \times \nabla)_i = (\mathbf{x} \times \mathbf{p})_i.$$

From Eq. (18.6) one recovers the Lorentz commutation relations; for example,

$$[J_i, J_j] = i\epsilon^{ijk}J_k.$$

18.2.2 Translations

We now complete the algebra by introducing spacetime translations, which together with Lorentz transformations form the full Poincaré group. Consider an infinitesimal translation generated by a constant four-vector ε_μ:

$$\tilde{x}_\mu = x_\mu - \varepsilon_\mu.$$

This is generated by the four-momentum operator

$$P_\mu \equiv i\,\partial_\mu = (i\partial_0;\ -i\nabla) \tag{18.8}$$

Indeed,

$$\tilde{x}_\mu = \exp(i\varepsilon_\rho P^\rho)x_\mu. \tag{18.9}$$

because:

$$(1 + i\,\varepsilon_\rho P^\rho)\,x_\mu = x_\mu - \varepsilon_\rho\,\delta^\rho{}_\mu = x_\mu - \varepsilon_\mu$$

where we have used:

$$i\,(i\partial^\rho)\,x_\mu = -\,\delta^\rho{}_\mu \qquad \text{with} \qquad \partial^\rho(x_\mu) = \delta^\rho{}_\mu.$$

Therefore the four-momentum operator generates spacetime translations.

The translation operators clearly commute with each (if I perform a translation in one direction followed by a translation in a different direction, if I were to invert the order of the two operations I end up with the same result):

$$[P_\mu, P_\nu] = 0$$

Also the commutator between the translation operator and the generalized angular momentum operator can be easily computed using the above relations, by direct computation:

$$[P_\mu, L_{\rho\sigma}] = i(\eta_{\mu\rho}P_\sigma - \eta_{\mu\sigma}P_\rho).$$

18.3 The Full Poincaré Algebra

A generic Poincaré transformation can be written as

$$\tilde{x}_\mu = \exp(i\alpha_\rho P^\rho)\exp\left(\tfrac{1}{2}i\epsilon_{\rho\sigma}L^{\rho\sigma}\right)x_\mu. \tag{18.10}$$

Collecting the commutation relations derived above, we obtain the complete Poincaré algebra, which is given by the following set of commutation relations:

$$[L_{\mu\nu}, L_{\rho\sigma}] = -i(\eta_{\mu\rho}L_{\nu\sigma} - \eta_{\mu\sigma}L_{\nu\rho} + \eta_{\nu\sigma}L_{\mu\rho} - \eta_{\nu\rho}L_{\mu\sigma}), \tag{18.11}$$

$$[P_\mu, P_\nu] = 0, \tag{18.12}$$

$$[P_\mu, L_{\rho\sigma}] = i(\eta_{\mu\rho}P_\sigma - \eta_{\mu\sigma}P_\rho). \tag{18.13}$$

Using the notation with Latin indices introduced earlier, the above relations can conveniently be expressed in terms of rotations $J_i = \frac{1}{2}\,\varepsilon_{ijk}\,L_{jk}$ and boosts $K_i = L_{0i}$ in the following way:

$$[P_0, J_i] = 0 \tag{18.14}$$

$$[P_i, J_j] = i\,\varepsilon_{ijk}\,P_k \tag{18.15}$$

$$[P_0, K_i] = i\, P_i \tag{18.16}$$

$$[P_i, K_j] = i\, P_0\, \delta_{ij} \tag{18.17}$$

One should notice that P_0 is a scalar under rotations, but P_i is a vector under rotations. Under a boost along the i-axis, both P_0 and P_i are affected, but the other two spatial components remain unchanged.

Finally, the Poincaré group can be written formally as

$$\mathrm{ISO}(3, 1) = \mathbb{R}^{3,1} \rtimes \mathrm{SO}(3, 1) \tag{18.18}$$

i.e. as a semidirect product between the group of continuous translations in Minkowski space-time, $\mathbb{R}^{3,1}$, and the Lorentz group. Here one should recall the definition of semidirect product of two groups that was given in Chap. 13, Sects. 13.5.3 and 13.6. To be even more precise, one can choose to restrict the Lorentz group to only *orthochronous* transformations, i.e. only those which preserve the direction of time, corresponding to the condition $\Lambda^0_0 \geq 1$ such that future-directed time-like vectors remain future-directed and transformations implying time-reversal, $t \to -t$, are excluded. This defines the orthochronous Lorentz group, denoted as $\mathrm{SO}^+(3, 1)$, and the corresponding orthochronous Poincaré group is given by $\mathrm{ISO}(3, 1) = \mathbb{R}^{3,1} \rtimes \mathrm{SO}^+(3, 1)$.

18.4 Casimir Invariants

Having established the structure of the Poincaré algebra, we now ask how its irreducible representations are classified. As in the case of other Lie algebras, the key objects are its Casimir invariants. The concept was first introduced by Dutch physicist Hendrik Casimir in 1931, in the context of his work on the quantum mechanics of a rotating rigid body.

The Poincaré algebra admits two Casimir invariants. The first is

$$P^2 = P_\mu P^\mu,$$

and it is easy to verify that, indeed,

$$[P^2, P_\mu] = 0, \qquad [P^2, L_{\rho\sigma}] = 0.$$

This Casimir invariant corresponds, physically, to the squared mass of the elementary particle, because, in special relativity, $P_\mu P^\mu = m^2$. Promoting P^2 to a quantum operator, the squared mass is its eigenvalue, hence the physical observable associated with it: $P^2|\psi\rangle = m^2|\psi\rangle$. Obviously, this also is consistent with the mass of the particle being Lorentz-invariant and the same in all inertial reference frames.

The second Casimir invariant is associated with the Pauli–Lubanski vector

$$W_\mu \equiv \frac{1}{2}\epsilon_{\mu\nu\rho\sigma} P^\sigma L^{\nu\rho}. \tag{18.19}$$

This object plays a central role in the representation theory of the Poincaré group, as it encodes the intrinsic angular momentum (spin) of the particle.

The corresponding Casimir invariant is $W^2 = W_\mu W^\mu$ and, indeed, it commutes with the full Poincaré algebra:

$$[W^2, P_\mu] = 0$$
$$[W^2, L_{\rho\sigma}] = 0.$$

The physical meaning of the Pauli-Lubanski pseudo-vector becomes transparent upon considering the rest-frame of the particle:

$$P_\mu \;\rightarrow\; \bar{P}_\mu \equiv (m;\ \mathbf{0})$$

in natural units $c = 1$, and the operator W_μ reduces to

$$W_0 = 0, \qquad W_i = \frac{1}{2}\, m\, \varepsilon_{ijk}\, L_{jk} = m\, J_i$$

This can be seen by writing the generalized angular momentum operator in terms of K and J by means of the relations

$$K_i = L_{0i}$$

$$J_i = \frac{1}{2}\varepsilon_{ijk}\, L_{jk}$$

from which the components of the Pauli–Lubanski vector follows as:

$$W_0 = \mathbf{J}\cdot\mathbf{P} = J_\mu P^\mu$$

$$\mathbf{W} = \mathbf{P}\wedge\mathbf{K} + E\,\mathbf{J} = \mathbf{P}\wedge\mathbf{K} + m\,\mathbf{J}$$

where, in natural units with $c = 1$, $E = p^0 = m$. From this observation we can see that the spatial components of the Pauli–Lubanski vector coincide, in the rest frame, with the angular momentum measured in the rest frame, that is, with the spin of the particle, Because, in the rest frame $P_{1,2,3} = 0$, then

$$\mathbf{W} = m\,\mathbf{J} \qquad \text{since} \qquad \mathbf{P} = 0.$$

Thus, the Casimir invariants of the Poincaré algebra correspond physically to the mass and spin of elementary particles. The Pauli–Lubanski vector can also be interpreted geometrically using the dual of the antisymmetric tensor constructed from P_μ and $L_{\rho\sigma}$.

Exercise
Prove that the Pauli–Lubanski pseudovector does indeed commute with all the generators of the Poincaré algebra.

Solution
The Pauli–Lubanski vector is defined as

$$W^\mu \equiv \frac{1}{2}\,\varepsilon^{\mu\nu\rho\sigma} P_\nu L_{\rho\sigma},$$

where P_μ are the generators of translations and $L_{\rho\sigma}$ are the generators of Lorentz transformations.

Its square is

$$W^2 \equiv W_\mu W^\mu,$$

which is a Lorentz scalar.

The Poincaré algebra contains the relations

$$[P_\mu, P_\nu] = 0,$$

$$[L_{\rho\sigma}, P_\mu] = i(\eta_{\mu\rho} P_\sigma - \eta_{\mu\sigma} P_\rho).$$

Consider the commutator with W^μ:

$$[P_\alpha, W^\mu] = \frac{1}{2}\varepsilon^{\mu\nu\rho\sigma} P_\nu [P_\alpha, L_{\rho\sigma}].$$

Using the Leibniz rule,

$$[P_\alpha, P_\nu L_{\rho\sigma}] = [P_\alpha, P_\nu]L_{\rho\sigma} + P_\nu[P_\alpha, L_{\rho\sigma}].$$

The first term vanishes since $[P_\alpha, P_\nu] = 0$. The second term vanishes because the contraction of the antisymmetric Levi–Civita tensor with a symmetric index structure is zero. Therefore,

$$[P_\alpha, W^\mu] = 0.$$

It follows immediately that

$$[W^2, P_\mu] = [W_\nu W^\nu, P_\mu] = 0.$$

The Pauli–Lubanski vector transforms as a Lorentz four-vector:

$$[L_{\rho\sigma}, W^\mu] = i(\eta^\mu_{\ \rho} W_\sigma - \eta^\mu_{\ \sigma} W_\rho).$$

Since $W^2 = W_\mu W^\mu$ is the contraction of a vector with itself, it is a Lorentz scalar. Scalars are invariant under Lorentz transformations, hence

$$[W^2, L_{\rho\sigma}] = 0.$$

The operator W^2 commutes with all generators of the Poincaré algebra:

$$[W^2, P_\mu] = 0, \qquad [W^2, L_{\rho\sigma}] = 0.$$

Therefore W^2 is a Casimir operator of the Poincaré group. Together with P^2, it labels the irreducible representations and characterizes the intrinsic spin (or helicity) of particles.

The Poincaré group therefore classifies elementary particles by mass and spin. However, these quantum numbers do not exhaust the full set of symmetries observed in nature, as we shall see in the next section.

18.5 The Symmetry of the Standard Model of Elementary Particles

The Poincaré group thus describes the basic symmetries of individual massive particles travelling at relativistic speeds in empty space. Mass and spin conservation do not, of course, exhaust the symmetries of elementary particles, which also include electric charge, weak isospin and colour charge, all of which are independent of the Poincaré group.

In particular, the conservation of electric charge follows from a global U(1) phase symmetry of the wave function:

$$\psi(x) \rightarrow e^{i\alpha}\psi(x),$$

where $\alpha \in \mathbb{R}$ is an arbitrary phase meaning that the physics does not depend on its value. By applying Noether's theorem [3]:

$$\text{U(1) symmetry} \longrightarrow \text{conserved current} \longrightarrow \text{conserved electric charge}$$

This U(1) symmetry is at the basis of electromagnetism: when it is made local, the electromagnetic field and the electromagnetic potential emerge. It is the same symmetry that is operative in classical covariant electrodynamics, where it corresponds to the gauge symmetry of the electromagnetic four-potential. Indeed, the four-potential is invariant under the addition of the derivative of a scalar field:

$$A_\mu \;\rightarrow\; A_\mu + \partial_\mu \Lambda$$

where $\Lambda(x)$ is a real scalar field defined over spacetime. Adding the symmetries of the weak force (weak isospin), SU(2), and of the strong force (colour charge) SU(3), we can write the total symmetry of the standard model of elementary particles as the following direct product of groups:

$$\text{Poincaré} \times \text{SU(3)} \times \text{SU(2)} \times \text{U(1)}$$

This symmetry group represents the hitherto unchallenged description of the smallest building blocks of matter.

References

1. H.F. Jones, *Groups, Representations and Physics*, 2nd edn. (Taylor & Francis Ltd, Bristol/New York, 1998)
2. A. Zee, *Group Theory in a Nutshell for Physicists* (Princeton University Press, Princeton, 2016)
3. L.D. Landau, E.M. Lifshitz, *The Classical Theory of Fields*. Course of Theoretical Physics, vol. 2, 4th edn. (Butterworth-Heinemann, Oxford, 1975)

Index

A. Zaccone, *Differential Geometry and Group Theory*,
https://doi.org/10.1007/978-3-032-27518-9

The manufacturer's authorised representative in the EU is Springer Nature Customer Service Centre GmbH, Europaplatz 3, 69115 Heidelberg, Germany. If you have any concerns regarding our products, please contact ProductSafety@springernature.com

Printed and bound by CPI Group (UK) Ltd, Croydon, CR0 4YY

05/08/2026

02186827-0002